T0341299

Advanced Optimization for Motion Control Systems

Advanced Optimization for Motion Control Systems

Jun Ma
Xiaocong Li
Kok Kiong Tan

CRC Press is an imprint of the
Taylor & Francis Group, an **informa** business

CRC Press
Taylor & Francis Group
6000 Broken Sound Parkway NW, Suite 300
Boca Raton, FL 33487-2742

Printed on acid-free paper

International Standard Book Number-13: 978-0-367-34339-2 (Hardback)

Library of Congress Control Number: 2020930274

**Visit the Taylor & Francis Web site at
http://www.taylorandfrancis.com**

**and the CRC Press Web site at
http://www.crcpress.com**

To our families

Contents

Preface

Motion control has gathered considerable momentum and research attention for more than a century. Continuous technological advancement in motion control enables various industries to achieve higher precision, product quality, and process safety, with less manufacturing costs and shorter manufacturing duration.

It is worth pointing out that although the research in motion control has progressed by leaps and bounds, it is still challenging to fulfill the ever-increasingly stringent performance standards. First, control theories are not easy to be directly applied to solve practical control problems, despite the remarkable theoretical advances and breakthroughs. On the other hand, the system performance is limited by practical issues, such as disturbances, noises, and perturbations, which render significant challenges to the design and implementation of control systems.

Substantial work has been conducted and reported in the core area of applying optimization approaches in motion control systems, which is a promising trend to achieve high performance for the next generation of industrial systems. In this book, various key issues with regard to optimization for motion control systems are addressed. The ultimate objective of the book is to provide readers with fundamental concepts and generic guidelines in control system designs through an optimization framework. The optimization can be done at a single manufacturing process level, or it can be enterprise-wide across a number of such processes to achieve improved performance of the plant. The Industry 4.0 framework can be leveraged to achieve such plant-wide optimization, tapping into IoT and cloud computing.

The book begins with an introduction to motion control systems and a brief survey of the model-based and the data-based optimization approaches. It is organized into two parts. Part I focuses on the model-based approaches including constrained linear quadratic optimization, constrained $\mathscr{H}_2$ optimization, and constrained $\mathscr{H}_2$ guaranteed cost optimization. Part II presents the data-based approaches, where reduced-order inverse model optimization, reference profile alteration and optimization, and disturbance observer sensitivity shaping optimization are addressed. To illustrate the practical appeal of the proposed optimization techniques, theoretical results are verified with practical examples in each chapter. Industrial problems explored in the book are formulated systematically with necessary analysis of the control system synthesis.

By virtue of the design and implementation nature, this book can be used as a reference for engineers, researchers, and students who want to utilize control theories to solve practical control problems. As the methodologies have extensive applicability in many control engineering problems, the research results in the field of optimization can be applied to full-fledged industrial processes, filling in the gap between research and application to achieve an improvement of frontier technologies. This book can also serve as a reference or textbook for a course at the graduate level.

This book is written with tremendous contributions from Prof. Tong Heng Lee, A/Prof. Abdullah Al Mamun, A/Prof. Arthur Tay, Dr. Si-Lu Chen, Dr. Chek Sing Teo, Dr. Nazir Kamaldin, and Dr. Wenyu Liang. The authors would like to express their gratitude and appreciation to them.

Berkeley, California — Jun Ma
Singapore — Xiaocong Li
Singapore — Kok Kiong Tan

Authors

Jun Ma received his B.Eng. (1st Class Hons.) degree in electrical and electronic engineering from the Nanyang Technological University, Singapore, in 2014, and his Ph.D. degree in electrical and computer engineering from the National University of Singapore, Singapore, in 2018. From 2018 to 2019, he was a Research Fellow with the Department of Electrical and Computer Engineering, National University of Singapore, Singapore. In 2019, he was a Research Associate with the Department of Electronic and Electrical Engineering, University College London, London, UK. He is currently a Visiting Scholar with the Department of Mechanical Engineering, University of California, Berkeley, Berkeley, California, USA. His research interests include control and optimization, precision mechatronics, robotics, and medical technology. He was a recipient of the Singapore Commonwealth Fellowship in Innovation.

Xiaocong Li received his B.Eng. and Ph.D. degrees in electrical engineering from the National University of Singapore, Singapore, in 2013 and 2017, respectively. He is currently a Research Scientist with the Mechatronics Group, Singapore Institute of Manufacturing Technology, Agency for Science, Technology and Research, Singapore. His research interests include precision motion control, data-driven intelligent control, and industrial automation.

Kok Kiong Tan received his B.Eng. (1st Class Hons.) and Ph.D. degrees in electrical engineering from the National University of Singapore, Singapore, in 1992 and 1995, respectively. Prior to joining the National University of Singapore, he was a Research Fellow at the Singapore Institute of Manufacturing Technology, Agency for Science, Technology and Research, a national R&D institute spearheading the promotion of R&D in local manufacturing industries, where he was involved in managing industrial projects. He is currently a Professor with the Department of Electrical and Computer Engineering, National University of Singapore, Singapore. He has authored or coauthored more than 200 journal papers to date and has written 14 books, all resulting from research in these areas. He has attracted research funding in excess of S$18 million to date and has received several research awards. His current research interests include precision motion control and instrumentation, advanced process control and auto-tuning, and general industrial automation.

Nomenclature

ARE	Algebraic Riccati Equation
ARMAX	Autoregressive Moving Average with Exogenous Inputs
CBT	Correlation-Based Tuning
CNC	Computer Numerical Control
CT	Computed Tomography
DHG	Dual-Drive H-Gantry
DOB	Disturbance Observer
DOF	Degree of Freedom
DSP	Digital Signal Processing
FDT	Frequency Domain Tuning
GA	Genetic Algorithm
IFT	Iterative Feedback Tuning
ILC	Iterative Learning Control
KKT	Karush-Kuhn-Tucker
LMI	Linear Matrix Inequality
LQG	Linear Quadratic Gaussian
LQR	Linear Quadratic Regulator
LTI	Linear Time-Invariant
MaxAE	Maximum Absolute Error
MFAC	Model-Free Adaptive Control
PAP	Profile Alteration and Prediction
PID	Proportional-Integral-Derivative
RMSE	Root Mean Squared Error
SISO	Single-Input-Single-Output
VRFT	Virtual Reference Feedback Tuning

1

Introduction

1.1 Overview of Motion Control Systems

Motion control is a sub-field of automation entering an era of rapid changes and technological advances. It encompasses the systems related to moving parts of machines in a controlled manner. The motion control system is widely used in various fields in order to develop automated systems, such as precision engineering, micromanufacturing, biotechnology, and nanotechnology (Tan, Lee, and Huang 2007). The main components involved in a motion control system include the motion controller, the motor drive, the motor, the encoder, as well as other mechanical components. Each of these plays a unique role in achieving precision motion control. The motion controller is the brain of the system controlling the motion path, the servo loop closure, and the sequence execution. The controller sends a low-power command signal to the motor drive in digital or analog form. The motor drive amplifies the signal, produces the torque and sets the load into motion. Finally, the feedback sensors record the performance and send information to the controller.

Although great achievements have been made in control engineering, it is still challenging to fulfill high performance standards for precision motion control systems. This is because the system performance is limited by practical issues, such as disturbances, noises and perturbations. To improve the characteristics of the control system, various control schemes can be implemented based on the requirements and specifications; these control schemes include classical control (Åström and Hägglund 1995; Tan, Wang, and Hang 2012; Wang, Lee, Fung, Bi, and Zhang 1999), optimal control (Anderson and Moore 1971; Burl 1999; Anderson and Moore 2007), robust control (Bhattacharyya and Keel 1994; Wang, Xie, and de Souza 1992; Xie 1996; Xie and de Souza Carlos 1992), nonlinear control (Rugh and Shamma 2000; Zhao, Tomizuka, and Isaka 1993; Utkin, Guldner, and Shi 2009) and intelligent control (Bristow, Tharayil, and Alleyne 2006; Xu and Tan 2003; Åström and Wittenmark 2013; Jang, Sun, and Mizutani 1997), etc. Selecting the best motion control scheme and controller parameters with respect to the system architecture is crucial as it largely determines the performance of the machine or the automated system. In most situations, the above control schemes can be integrated, and the controller design problem can be converted to a multi-objective optimization problem with certain design requirements.

1.2 Optimization Methods

Optimization of a collection of objectives systematically and simultaneously is a multi-objective optimization problem, where these objectives are the criteria of different design targets. These design targets could be incommensurable and competing, and it becomes a burden to find the optimal solution with trade-off surfaces. Classical methods to obtain the optimal solution will aggregate multiple objectives into a single form parameterized objective function, such as the weighting method and the constraint method. The weighting method converts a multi-objective optimization problem into a single optimization problem by forming a linear combination of the objectives. However, this approach is not capable of deriving the optimal solution with non-convex trade-off surfaces. On the other hand, the constraint method transforms multiple objectives into constraints. Remarkably, the constraint method is suitable for the case where a prior preference on the multiple objectives is available.

In the literature, multi-objective optimization not only plays a significant role in motion control systems but also has extensive applicability in many manufacturing processes. A multi-objective optimization approach is applied to process planning in (Sheng, Srinivasan, and Kobayashi 1995). Based on this approach, the optimization model is further extended in cutting tool parameters planning and process sequence selection. In (Rajemi, Mativenga, and Aramcharoen 2010), turning conditions are optimized for cost and energy objectives. Recently, research on process energy analysis based on statistical or artificial intelligence approaches has been conducted. For example, in (Cho, Park, Choi, and Leu 2000), the trade-off between dimensional error and laser power is analyzed by using GA. Also, a Taguchi-based statistical approach is applied to understand the effect of laser power on photo-polymers in the stereolithography process (Campanelli, Cardano, Giannoccaro, Ludovico, and Bohez 2007). In (Paul and Anand 2012), a generic methodology is developed to analyze the part shape and the energy consumption of the selective laser sintering process. It is declared that global optimum is crucially important for industrial optimization problems, but the methods proposed by recent research are limited.

The solutions of a mathematically formulated optimization problem together with optimization methods allow a considerable number of different design variants to be calculated. They also make it possible to perform these calculations at production planning stages for a prototype to possess the qualities given by a chosen criteria function. To solve the optimization problem for a motion control system, the following procedures are required:

- Formulation of a mathematical description
- Specification of the objective function
- Selection of the optimization variables

- Specification of the constraining functions
- Selection of a suitable optimization method
- Solving of the formulated optimization problem
- Transformation of the optimization results back into the dynamic model

Various types of optimization methods can be used to solve an optimization problem (Bishop 2005). Generally, these methods can be categorized into two types: standard optimization methods and evolutional optimization methods.

Standard optimization methods can be categorized according to the order of the derivatives used in an optimization approach. Some of these methods are listed below.

- Zero-order methods

  Golden-section search (Nazareth and Tseng 2002): Golden-section search aims to minimize a unimodal continuous function over an interval without using derivatives, by successively narrowing the range of values inside which the minimum exists. Golden-section search derives its name because the algorithm maintains the function values for triples of points, whose distances form a golden ratio 1.618.

  Simplex method (Nelder and Mead 1965): Simplex method is a technique to solve problems in linear programming. This method was first invented by George Dantzig in 1947 (Dantzig 2016), to test adjacent vertices of the feasible set in sequence such that the objective function improves or remains unchanged at each new vertex.

  Stochastic method (Boender, Kan, Timmer, and Stougie 1982): Stochastic method requires the calculation of the objective function values at a large number of selected points, where all of the points have equal probability to be selected in the space. It follows that the stochastic method results in the computation of the function values at many points, which may protract the calculation.

- First-order methods

  Bisection method (Byers 1988): Bisection method is also called the interval halving method, the binary search method or the dichotomy method. By definition, it is a root-finding method that repeatedly bisects an interval and then selects a subinterval in which a root must lie for further processing. Bisection method has the advantage in terms of simplicity and robustness, but the speed of this method is slow. Therefore, this method is often used to obtain an approximation to a solution, and then the optimization result is used as a starting point for more rapidly converging algorithms.

  Conjugate gradient method (Powell 1977): Conjugate gradient method is an algorithm to seek the nearest local minimum of an objective function which presupposes that the gradient of the function can be computed, but it uses conjugate directions instead of the local gradient for progressing.

Variable-metric method (Davidon 1991): Variable-metric method, also known as the quasi-Newton method, is targeted to accumulate information from successive line minimizations. By using this method, the quadratic convergence is guaranteed for general smooth functions.

- Second-order method:

  Newton's method (Bertsekas 1999): It is also called the Newton-Raphson method, which is a root-finding algorithm that uses first and second derivatives. Providing a starting point, a quadratic approximation to the objective function is constructed to match the first and second derivative values at that point, then the approximate function is minimized, and the minimizer of the approximate function is iteratively used as the starting point in the next iteration.

However, not all problems can be easily solved by standard optimization methods, so a number of evolutional optimization methods are designed based on the laws of some natural processes. These heuristic approaches are generally very well specified, so they can be widely applied. Moreover, they are able to get out of the trap of a local minimum. Despite the advantage of heuristic approaches, it is not guaranteed that the global optimum can be reached when the optimization process is terminated. Some typical heuristic algorithms are listed below.

Stochastic hill climbing algorithm (Mondal, Dasgupta, and Dutta 2012): Stochastic hill climbing algorithm is a variant of the gradient method, but the steepest climbing direction is determined by searching the neighborhood. Similar to the gradient method, this algorithm is likely to trap in a local minimum. Since the problem of looping exists by using such an algorithm, the algorithm needs to be executed several times with a variety of random initial values, and then the best result among them is chosen.

Tabu search algorithm (Cordeau, Gendreau, and Laporte 1997): In 1986, Tabu search algorithm was created by Professor Fred W. Glover to seek the global minimum. As compared to the hill climbing algorithm, looping problems caused by the trap of a local minimum can be successfully eliminated due to the use of short-time memory.

Simulated annealing algorithm (Corana, Marchesi, Martini, and Ridella 1987): The name of the simulated annealing algorithm comes from the fact that it mimics the process undergone by misplaced atoms in metal when it is heated and then slowly cooled. This algorithm is very effective in solving some tough problems such as the traveling salesman problem, which belongs to the NP-complete class of problems.

GA (Deb, Pratap, Agarwal, and Meyarivan 2002): Inspired by the process of natural selection, GAs are most frequently used to optimize the parameters for a system with a complicated mathematical description or even unknown model (Lee and Hajela 1996), by relying on bio-inspired operators such as mutation, crossover, and selection. Apart from general optimization problems, GAs are useful in neural networks problems, either for seeking suitable weights or optimizing the structure of a neural network. When executing GAs, the function values for tens or hundreds of genetic chains in a population need to be calculated and a large number of populations during a single run of the program are required to be evaluated, so GAs are time-consuming. Therefore, GAs are not suitable to optimize relatively simple functions.

1.3 Model-Based Optimization for Motion Control Systems

Model-based optimization is an innovative optimization strategy, where the design environment facilitates general communication between various design factors. Through the construction of mathematical models of the plant using empirical equations, specification of the objective function and constraining functions, selection of the optimization variables, the model-based approach allows the use of various mathematical and numerical optimization methods. In certain scenarios, it can be trivial to determine the optimal policy with the task and the plant specified.

Optimal control theory already attracted considerable attention for many years due to its effectiveness in model-based control system synthesis problems. It aims to find a control law for a given system, such that a certain optimality criterion is achieved. Optimal control is closely related in its origins to the theory of calculus of variations (Goldstine 1981; Wan 1995; Leitmann 1981). There are some important milestones in the development of optimal control in the 20th century (Wilamowski and Irwin 2011), such as the formulation of dynamic programming by Richard Bellman (Bellman 1957; Bertsekas 1995), the development of the minimum principle by Lev Pontryagin in the 1950s (Pontryagin 1987), the formulation of the LQR and the Kalman filter by Rudolf Kalman in the 1960s (Anderson and Moore 1971).

An optimal control problem consists of a set of differential equations describing the paths of the control variables, and it aims to minimize a given cost function, which could be solved based on Pontryagin's maximum principle (Sussmann and Willems 1997; Kopp 1962), or by solving the Hamilton-Jacobi-Bellman equation (Bellman 1957). By such an indirect method, the optimal control problem is converted to a boundary-value problem, and the calculus of variations is used to determine the first-order optimality conditions of the problem (Rao 2009). However, the solution to a class of optimal control problems cannot be found by analytical means (Wilamowski and Irwin 2011). On the other hand, direct methods transform the original optimal control problem via discretization of the control and the state functions on a time grid to a nonlinear constrained optimization problem or a nonlinear programming problem (Becerra 2004; Betts 2001; Bazaraa, Sherali, and Shetty 2005; Böhme and Frank 2017).

A special case of the optimal control problem which is of particular importance arises when the cost function is a quadratic function in terms of the state variables and the control inputs, where the dynamic equations are linear. The resulting feedback control law in this case is known as the LQR. Notably, there are well-established methods and software to solve the ARE for the LQR design (Roberts and Becerra 2000). An important extension of the LQR concept to systems with Gaussian additive noise is known as the LQG control (Bernstein and Haddad 1989). The LQG control involves coupling the LQR with the Kalman filter using the separation principle. Also, it has been widely applied in controller design problems.

For the analytical design of control systems, it is often convenient to measure the system performance in terms of the norm of the closed-loop system transmittance from the exogenous signals to the regulated variables (Chen and Saberi 1993; Chen, Saberi, Sannuti, and Shamash 1993). After choosing a suitable norm and synthesis model of the plant, the control system synthesis proceeds by finding a controller with the norm minimized. One common measure of performance for a linear system is the $\mathscr{H}_2$-norm. This norm is relevant in the minimum variance control problems, or the infinite-horizon stationary-statistics LQG problems (Doyle, Glover, Khargonekar, and Francis 1989). An $\mathscr{H}_2$ control problem targets stabilizing the control system while minimizing the $\mathscr{H}_2$-norm of the respective transfer function (Kučera 2007). The importance of the $\mathscr{H}_2$-norm is further emphasized by mathematical tractability and the standard $\mathscr{H}_2$ control problem has a closed-form solution by solving the AREs.

Although the feedback controller parameters can be easily optimized by direct computation for classical LQR or $\mathscr{H}_2$ problems via the use of the AREs, these methods cannot be employed directly when the gain matrix is under prescribed structural constraints. These constraints arise due to different reasons. The first reason is that certain elements of the problem (such as disturbances, computational capacity, reconfigurability, ease of cabling and tuning, etc.) are application-specific and they change between industrial environments (Bakule 2008; Dolk, Borgers, and Heemels 2017). Consequently, the controller structure is specially designed and implemented. For instance, in some seminal works, the design of an optimal decentralized controller subjected to prescribed sparsity pattern has been carried out (Wang, Matni, and Doyle 2018). Especially, for fully decentralized control systems in which measurement and control of subsystems are independent of each other, the off-diagonal elements in the gain matrix are constrained to be zero (Cvejn and Tvrdík 2017). The second reason is due to the possible incorporation of mechanical components. This is quite common in a class of integrated mechatronic system design problems (Bozca, Muğan, and Temeltaş 2008), where mechanical components naturally introduce certain structural constraints to the composite gain matrix (Ma, Chen, Kamaldin, Teo, Tay, Al Mamun, and Tan 2018; Ma, Chen, Kamaldin, Teo, Tay, Al Mamun, and Tan 2017; Ma 2017; Ma, Chen, Liang, Teo, Tay, Al Mamun, and Tan 2019; Ma, Chen, Teo, Tay, Al Mamun, and Tan 2019; Ma, Chen, Teo, Kong, Tay, Lin, and Al Mamun 2017), for example, the use of spring, flexure and damper (Tehrani, Jalaleddini, and Kearney 2017; Zhu, Pang, and Teo 2016). This is because to employ the classical optimal control approach to integrated mechatronic design problems, the composite gain matrix is derived through the augmentation of mechanical parameters and controller gains. To sum up, due to the specially designed control structure and the naturally imported constraints, two types of constraints are commonly addressed in motion control systems: (1) Certain elements in the composite gain matrix are forced to be zero; (2) Some elements in the gain matrix are equal or opposite. Indeed, these constraints make the optimization problem more challenging than its unconstrained centralized counterpart, and the constrained optimization problem becomes NP-hard (Tsitsiklis and Athans 1985; Fattahi and Lavaei 2017). Thus, it leaves an open problem on the design of controllers for motion control problems, to optimize a pre-defined performance index under the above two types of constraints.

1.4 Data-Based Optimization for Motion Control Systems

Industrial motion and process control systems are becoming more and more complex and it is often not an easy job to obtain the system model. Even though a plant model can be obtained by first principles or identification in a less complex system, it is sometimes not accurate enough for the controller design. In this case, the resulting inaccurate model-based controller design may not provide satisfactory performance. On the other hand, rich information is contained in the huge amount of input-output data during the actual operations. Although such information is easy to collect, it is often not used as a tool to further improve the control performance in a systematic way. For example, the control engineer may want to fine-tune an existing PID controller when the performance in terms of tracking or disturbance rejection is not satisfactory, but the tuning is often done based on the experience of the control engineer by trial-and-error.

In view of this fact, many data-based controller design methods have been developed. In contrast to the traditional model-based control, it makes direct use of the information contained in the input-output data without explicitly obtaining an accurate system model. With the additional information extracted from the input-output data, the control performance is expected to be improved further. Various promising data-based techniques in the motion control and process control context have been continuously developed by control researchers and engineers, and many of them have already been applied in the industry, e.g. the IFT (Hjalmarsson, Gevers, Gunnarsson, and Lequin 1998; Hjalmarsson 2002; Li 2017; Li, Chen, Teo, and Tan 2019; Tan, Li, Chen, Teo, and Lee 2019), VRFT (Campi and Savaresi 2006), ILC (Bristow, Tharayil, and Alleyne 2006), MFAC (Hou and Jin 2011b; Hou and Jin 2011a), FDT (Kammer, Bitmead, and Bartlett 2000), CBT (Karimi, Mišković, and Bonvin 2004), etc.

The objectives of many control problems can be defined in terms of a cost function, and the purpose is to reduce the cost function to its minimum, preferably its global minimum. The optimization of such cost function usually requires gradient-based minimization and the gradient computation is often the main problem to solve. Typically, the gradient is a rather complicated function of the process or motion dynamics, i.e. system model. Thus, if the model for the process or motion is not accurate enough, the optimization process cannot offer the real optimal solution. Hence, it is desirable to search for an optimization algorithm without relying on an accurate system model. The key step of iterative feedback tuning is to find out an unbiased estimation for the gradient of the cost function based solely on the signals from the closed-loop experiments. The data is collected from an existing motion or process control system that is already in operation in a stable but not optimal way. To speed up the optimization process, an approximation of the Hessian of the cost function can also be estimated from the collected data. With this optimization procedure, the controller parameter is updated in each iteration and then the new parameter value is used in the next iteration of experiments. This iterative optimization normally takes a few iterations to complete and the stopping criterion can be defined by the users, for example when the cost function reduction is less than a certain threshold. One of the drawbacks of this method is that only the local minimum can be obtained theoretically, so a proper choice of the initial parameter is preferable if the purpose is to find the global minimum.

Despite the success of data-based methods in the PID controller, a more complicated control structure is necessary when even higher performance is required. In tracking control applications, another traditional approach to further improve the control performance is by using a 2-DOF controller. In motion systems that require high tracking performance with short move time and settling time, this 2-DOF control structure is widely used in many commercial controllers. The feedback controller is used for disturbance rejection and maintaining system stability but it has a lag in transient tracking because the error has to occur first before the feedback controller makes any adjustment. Thus, the feedforward controller is employed to improve tracking performance. In addition, in many motion control systems, disturbances such as friction, backlash, and cogging forces need to be eliminated or at least attenuated in order to achieve high accuracy tracking. To deal with these low-frequency disturbances effectively, a DOB is included in the inner control loop in addition to the original 2-DOF controller. The DOB can attenuate external disturbances as well as disturbances resulting from the model mismatch, while the outer loop feedback controller can be designed based on the nominal model. The model-based design for this advanced control structure with feedforward and DOB is commonly used in many applications and similar to other model-based methods, its performance is highly dependent on modeling accuracy. Hence, in the second half of this book, we will focus on developing data-based optimization methods for this particular control structure to further enhance its performance.

Part I

Model-Based Optimization for Motion Control Systems

2

Constrained Linear Quadratic Optimization

2.1 Background

The LQR is an optimal control problem where the state equation of the plant is linear, the cost function is quadratic, the test conditions consist of initial conditions on the state, and there are no disturbance inputs. The quadratic cost function can be used in a wide range of applications by appropriately selecting the weightings on the state and the control input. The objective is to compute a state feedback controller

$$u(t) = Kx(t), \tag{2.1}$$

that stabilizes the closed-loop system and minimizes the cost function

$$J = \int_0^\infty (x(t)^T Q x(t) + u(t)^T R u(t))\, dt, \tag{2.2}$$

where $Q \geq 0$, $R > 0$, x and u are the state variable and the control input of the LTI system

$$\dot{x}(t) = Ax(t) + Bu(t), \tag{2.3}$$

with $x(0) = x_0$, and (A, B) is assumed to be stabilizable, $(A, \sqrt{Q})$ is assumed to be detectable.

The closed-loop cost is given by

$$J = \int_0^\infty x(t)^T (Q + K^T R K) x(t)\, dt, \tag{2.4}$$

and the closed-loop system is given by

$$\dot{x}(t) = (A + BK)x(t). \tag{2.5}$$

For a given K and x_0, the state equation is given by

$$x(t) = e^{(A+BK)t} x_0. \tag{2.6}$$

Hence,

$$J = x_0^T \left(\int_0^{\infty} e^{(A+BK)^T t}(Q + K^T RK)e^{(A+BK)t}\, dt \right) x_0. \tag{2.7}$$

Thus, the optimal cost J^* is computed as

$$J^* = x_0^T P x_0, \tag{2.8}$$

where P is the solution to the Lyapunov equation

$$(A + BK)^T P + P(A + BK) + Q + K^T RK = 0, \tag{2.9}$$

which can be rewritten as

$$\begin{aligned} A^T P + PA - PBR^{-1}B^T P + Q \\ +(PBR^{-1} + K^T)R(R^{-1}B^T P + K) = 0. \end{aligned} \tag{2.10}$$

Since K is confined to the term

$$(PBR^{-1} + K^T)R(R^{-1}B^T P + K) \geq 0, \tag{2.11}$$

the optimal gain is computed as

$$K = -R^{-1}B^T P, \tag{2.12}$$

where P is the solution to the following ARE

$$A^T P + PA - PBR^{-1}B^T P + Q = 0. \tag{2.13}$$

It is worth pointing out that LQR optimal control exhibits some important properties. First, the LQR generates a static gain matrix. Hence, the order of the closed-loop system is the same as that of the plant. Second, the LQR achieves infinite gain margin $k_g = \infty$, and it also guarantees phase margin $\gamma \geq 60$ degrees. This is in good agreement with the practical guidelines for control system design.

2.2 Constrained Linear Quadratic Optimization Algorithm

Given an LTI system (2.3) with a state feedback controller (2.1), where $x \in \mathbb{R}^n$ is the state vector, $u \in \mathbb{R}^m$ is the control input vector, $K \in \mathbb{R}^{m \times n}$ is the controller gain matrix, matrices $A \in \mathbb{R}^{n \times n}$, $B \in \mathbb{R}^{n \times m}$ are assumed to be known. The objective function is defined as (2.2) with $Q \geq 0$, $R > 0$. The objective is to minimize (2.2) with certain constraints on the gain matrix K. Notice that the equality constraints can be expressed as $K \in \Phi$, where $\Phi = \{K \in \mathbb{R}^{m \times n} : \mathcal{C}(K) = \mathcal{C}_0\}$, and $\mathcal{C}(K)$ is a function of K with $\mathcal{C}(0) = 0$. Thus, the equality constraints on K can be written as

$$\mathcal{C}_1(K) = \tilde{A}_1 K \tilde{B}_1 + \tilde{A}_2 K \tilde{B}_2 + \ldots + \tilde{A}_s K \tilde{B}_s = 0, \tag{2.14a}$$

$$\mathcal{C}_2(K) = \tilde{A}_{s+1} K \tilde{B}_{s+1} + \tilde{A}_{s+2} K \tilde{B}_{s+2} + \ldots + \tilde{A}_t K \tilde{B}_t = 0, \tag{2.14b}$$

$$\ldots$$

$$\mathcal{C}_N(K) = \tilde{A}_w K \tilde{B}_w + \tilde{A}_{w+1} K \tilde{B}_{w+1} + \ldots + \tilde{A}_z K \tilde{B}_z = 0, \tag{2.14c}$$

where N is the number of equality constraints.

In addition, the stability of the closed-loop system needs to be guaranteed. Thus, we have the inequality constraint

$$\mathrm{Re}(\mathrm{eig}(A_\mathrm{c})) < 0, \tag{2.15}$$

with $A_\mathrm{c} = A + BK$, $\mathrm{Re}(\cdot)$ representing the real part of a number, $\mathrm{eig}(\cdot)$ representing the eigenvalue of a matrix.

Meanwhile, the cost function (2.2) is equivalent to

$$J(K) = \mathrm{Tr}(P(K)X_0), \tag{2.16}$$

where $\mathrm{Tr}(\cdot)$ represents the trace of a matrix, $P(K) = \int_0^\infty \Lambda_\mathrm{c}(t)^T Q_\mathrm{c} \Lambda_\mathrm{c}(t)\, dt$, $X_0 = x_0 x_0^T$, $\Lambda_\mathrm{c}(t) = e^{A_\mathrm{c} t}$, $Q_\mathrm{c} = Q + K^T R K$, and x_0 is the vector corresponding to the initial value of the state x. Hence, the optimization problem targets minimizing $J(K)$ in (2.16) under the constraints (2.14) and (2.15).

Remark 2.1 *The optimal solution of the classical LQR problem is independent of initial state values. But this property is lost when the gain matrix is under certain structural constraints.*

Remark 2.2 *When initial values of all the states are not available for the optimization, x_0 can be assumed to be a uniformly distributed random variable over the surface of an n-dimensional unit sphere. In this case, minimizing the expected value of the cost function $J(K)$ in (2.16) over x_0 is equivalent to minimizing the cost $\tilde{J}(K)$, where $\tilde{J}(K) = \mathrm{Tr}(P(K))$. In fact, $\tilde{J}(K)/n$ is the average value of the cost $J(K)$ in (2.16) as x_0 varies over the surface of the unit sphere.*

When there are constraints in the gain matrix, there is no standard closed-form solution to this optimization problem; this is because the ARE for the standard LQR problem cannot be used. To solve this problem, a gradient-based constrained optimization algorithm is proposed.

First, if only the equality constraints in (2.14) are considered in the optimization problem, then the optimization problem becomes

$$\min_{K \in \Phi} J(K). \tag{2.17}$$

We assume that $K \in \Phi$ is defined such that the gradient matrix $dJ(K)/dK$ is not null, and the optimization problem (2.17) is converted to calculate the optimal projection gradient matrix D to minimize its distance with its unconstrained counterpart. In addition, the associated step length α along the D will guarantee that the functional cost is always decreasing throughout iterations.

In this book, $\| \cdot \|$ represents the Frobenius norm of a matrix. Then this problem can be expressed as

$$\min_{D} \frac{1}{2} \left\| \frac{dJ(K)}{dK} - D \right\|^2, \quad s.t. \mathcal{C}_i(D) = 0, \quad \forall i = 1, \dots, N. \tag{2.18}$$

Problem (2.18) is further converted to the dual problem as

$$\max_{\Lambda_i} \min_{D} \left(\frac{1}{2} \left\| \frac{dJ(K)}{dK} - D \right\|^2 + \mathrm{Tr}\left(\sum_{i=1}^{N} \Lambda_i^T \mathcal{C}_i(D) \right) \right), \tag{2.19}$$

where Λ_i is the dual variable associated with the equality constraint $\mathcal{C}_i(D) = 0$. In the coming algorithm, we aim to update the gain K from the constrained gradient D iteratively. To ensure the monotonically decreasing of the cost for every computed D, we have the following theorem.

Theorem 2.1 *The optimization result is such that*

1. *if* $\|D\| \neq 0, \exists \bar{\alpha} > 0, J(K - \alpha D) < J(K), \forall 0 < \alpha \leq \bar{\alpha}$.
2. *if* $\|D\| = 0$, *KKT conditions of optimization problem* (2.17) *are equivalent to those of* (2.18).

Proof of Theorem 2.1: To prove Theorem 2.1, the following lemma is introduced first.

Lemma 2.1 *((Geromel and Bernussou 1982)) Let* $f(X) : \mathbb{R}^{m \times n} \to \mathbb{R}$ *be a function such that* $f(X + \varepsilon \delta X) = f(X) + \varepsilon \mathrm{Tr}(M(X)\delta X)$ *for all* $\delta X \in \mathbb{R}^{m \times n}$ *when* $\varepsilon \to 0$, *then*

$$\frac{df(X)}{dX} = M(X)^T. \tag{2.20}$$

From Lemma 2.1, it is easy to obtain

$$\begin{aligned} J(K - \alpha D) &= J(K) - \alpha \left(\frac{dJ(K)}{dK} \right)^T D + O(\alpha^2) \\ &= J(K) - \alpha \mathrm{Tr}\left(\left(\frac{dJ(K)}{dK} \right)^T D \right) + O(\alpha^2), \end{aligned} \tag{2.21}$$

with $\lim_{\alpha \to 0} O(\alpha^2)/\alpha = 0$. $D = 0$ is always feasible in problem (2.18), we have

$$\left\| \frac{dJ(K)}{dK} - D \right\|^2 \leq \left\| \frac{dJ(K)}{dK} \right\|^2. \tag{2.22}$$

So

$$\mathrm{Tr}\left(\left(\frac{dJ(K)}{dK}-D\right)^T\left(\frac{dJ(K)}{dK}-D\right)\right)\leq\mathrm{Tr}\left(\left(\frac{dJ(K)}{dK}\right)^T\left(\frac{dJ(K)}{dK}\right)\right),\tag{2.23}$$

which leads to

$$\mathrm{Tr}\left(\left(\frac{dJ(K)}{dK}\right)^T D\right)\geq\frac{1}{2}\mathrm{Tr}(D^T D)=\frac{1}{2}\left\|D\right\|^2.\tag{2.24}$$

Therefore,

$$J(K-\alpha D)-J(K)=-\alpha\mathrm{Tr}\left(\left(\frac{dJ(K)}{dK}\right)^T D\right)\leq-\frac{1}{2}\alpha\left\|D\right\|^2,\tag{2.25}$$

with $\|D\|\neq 0$. This proves Statement 1 of Theorem 2.1.

Consider the necessary optimal conditions for (2.17) and we have

$$\frac{dJ(K)}{dK}+\sum_{j=1}^{s}\tilde{A}_j^T\Gamma_1\tilde{B}_j^T+\sum_{j=s+1}^{t}\tilde{A}_j^T\Gamma_2\tilde{B}_j^T+\ldots+\sum_{j=w}^{z}\tilde{A}_j^T\Gamma_N\tilde{B}_j^T=0,\tag{2.26}$$

$$K\in\Phi,\tag{2.27}$$

$$\sum_{j=1}^{z}\tilde{A}_jK\tilde{B}_j=0.\tag{2.28}$$

Similarly, for the projection problem, we have

$$D-\frac{dJ(K)}{dK}+\sum_{j=1}^{s}\tilde{A}_j^T\Lambda_1\tilde{B}_j^T+\sum_{j=s+1}^{t}\tilde{A}_j^T\Lambda_2\tilde{B}_j^T+\ldots+\sum_{j=w}^{z}\tilde{A}_j^T\Lambda_N\tilde{B}_j^T=0,\tag{2.29}$$

$$D\in\Phi,\tag{2.30}$$

$$\sum_{j=1}^{z}\tilde{A}_jD\tilde{B}_j=0.\tag{2.31}$$

If $\|D\|=0$, the original problem is equal to the projection problem with $\Gamma_i=-\Lambda_i$, where $i=1,\ldots,N$. This proves Statement 2 of Theorem 2.1.

To compute the gradient matrix projected onto the equality constrained hyperplane, we have the following theorems.

Theorem 2.2 *For the finite horizon case ($T<\infty$) in the LQR problem, the matrical gradient of the functional cost with respect to the gain matrix is given by*

$$\frac{dJ(K)}{dK}=2\int_0^T(RK+B^TP(t))X(t)\,dt,\tag{2.32}$$

with the following two stationarity conditions

$$A_c^T P(t) + P(t)A_c + Q_c = -\dot{P}(t), \tag{2.33}$$

$$A_c X(t) + X(t)A_c^T = \dot{X}(t). \tag{2.34}$$

Proof of Theorem 2.2: The following lemma is given first.

Lemma 2.2 *(Geromel 1979) Define a general cost function and associated dynamical system as*

$$J(K) = \int_0^T f(X(t), K)\, dt + g(X(T)), \tag{2.35}$$

$$\dot{X}(t) = F(X(t), K), \quad X(0) = X_0, \tag{2.36}$$

where $X(t) \in \mathbb{R}^{n\times n}$ is the state variable matrix and $K \in \mathbb{R}^{m\times r}$ is the gain matrix. If the functions f, g and F are differentiable, the matrical gradient of $J(K)$ is given by

$$\frac{dJ(K)}{dK} = \int_0^T \frac{\partial}{\partial K} H(X(t), P(t), K)\, dt, \tag{2.37}$$

where

$$H(X(t), P(t), K) = f(X(t), K) + Tr(P(t)^T F(X(t), K)). \tag{2.38}$$

$P(t)$ and $X(t)$ are the solutions of

$$\frac{\partial H(X(t), P(t), K)}{\partial X(t)} = -\dot{P}(t), \quad P(T) = \frac{\partial g(X(T))}{\partial X(T)},$$

$$\frac{\partial H(X(t), P(t), K)}{\partial P(t)} = \dot{X}(t), \quad X(0) = X_0. \tag{2.39}$$

First, we rewrite the cost function in (2.16) as

$$J(K) = \int_0^T f(X(t), K)\, dt, \tag{2.40}$$

where

$$f(X(t), K) = \mathrm{Tr}(Q_c X(t)), \tag{2.41}$$

$$\dot{X}(t) = A_c X(t) + X(t)A_c^T. \tag{2.42}$$

From (2.38), the Hamiltonian function is

$$H(X(t), P(t), K) = \mathrm{Tr}(P(t)(A_c X(t) + X(t)A_c^T) + Q_c X(t)). \tag{2.43}$$

Notice that

$$\frac{\partial}{\partial K}\mathrm{Tr}(Q_{\mathrm{c}}X(t)) = 2RKX(t), \tag{2.44}$$

$$\frac{\partial}{\partial K}\mathrm{Tr}(P(t)(A_{\mathrm{c}}X(t) + X(t)A_{\mathrm{c}}^T)) = 2(P(t)B)^T X(t)^T. \tag{2.45}$$

From (2.37), (2.44) and (2.45) , the matrical gradient of the functional cost is given by (2.32) with two stationarity conditions (2.33) and (2.34). This concludes the proof of Theorem 2.2.

Theorem 2.3 *For the infinite horizon case ($T = \infty$) in the LQR problem, the matrical gradient of the functional cost with respect to the gain matrix is given by*

$$\frac{dJ(K)}{dK} = 2(RK + B^T P)V; \tag{2.46}$$

P and V are solutions of the two Lyapunov equations

$$A_{\mathrm{c}}^T P + PA_{\mathrm{c}} + Q_{\mathrm{c}} = 0, \tag{2.47}$$

$$A_{\mathrm{c}}V + VA_{\mathrm{c}}^T + X_0 = 0. \tag{2.48}$$

Proof of Theorem 2.3: It is easy to prove that for the infinite horizon case ($T = \infty$), (2.32) can be simplified to (2.46). Boundary conditions (2.33) and (2.34) are simplified to (2.47) and (2.48). This concludes the proof of Theorem 2.3.

Theorem 2.4 *The projection gradient matrix onto the equality constrained hyperplane is given by*

$$D = \frac{dJ(K)}{dK} - \sum_{j=1}^{s} \tilde{A}_j^T \Lambda_1 \tilde{B}_j^T - \sum_{j=s+1}^{t} \tilde{A}_j^T \Lambda_2 \tilde{B}_j^T - \ldots - \sum_{j=w}^{z} \tilde{A}_j^T \Lambda_N \tilde{B}_j^T. \tag{2.49}$$

Proof of Theorem 2.4: The necessary and sufficient optimal solution to the dual problem (2.19) is given by

$$D - \frac{dJ(K)}{dK} + \frac{\partial}{\partial D}\left(\mathrm{Tr}\left(\sum_{i=1}^{N} \Lambda_i^T C_i(D)\right)\right) = 0. \tag{2.50}$$

Thus,

$$D - \frac{dJ(K)}{dK} + \sum_{j=1}^{s} \tilde{A}_j^T \Lambda_1 \tilde{B}_j^T + \sum_{j=s+1}^{t} \tilde{A}_j^T \Lambda_2 \tilde{B}_j^T + \ldots + \sum_{j=w}^{z} \tilde{A}_j^T \Lambda_N \tilde{B}_j^T = 0, \tag{2.51}$$

which gives (2.49). This concludes the proof of Theorem 2.4.

Algorithm 2.1 Constrained Linear Quadratic Optimization Algorithm

- Step 1: Set $i = 0$ and initialize a stable gain $K^0 \in \Phi$.
- Step 2: Set $i = i + 1$ and determine the unconstrained gradient matrix $(dJ(K)/dK)^i$ by Theorem 2.3.
- Step 3: Determine the projection gradient matrix D^i by Theorem 2.4.
- Step 4: Optimize the step size α^i to minimize the functional cost after the iteration, which is a line search problem written as $\min\limits_{\alpha^i} J(K^{i-1} - \alpha^i D^i)$.
- Step 5: Gain matrix is updated as $K^i = K^{i-1} - \alpha^i D^i$.
- Step 6: Go back to Step 2 to continue the iterations until reaching the stopping criterion.

The targeted optimization problem (2.17) is a non-convex constrained optimization problem, while it can be converted to a convex optimization problem defined by (2.18). To solve the dual problem (2.19), we propose the following gradient-based optimization algorithm.

The following theorem is used to show that under certain conditions, both the equality constraints in (2.14) and the inequality constraint in (2.15) are satisfied for optimized K^i for all $i \geq 1$.

In this book, I represents an identity matrix with appropriate dimensions, and I_n represents an identity matrix with a dimension of $n \times n$.

Theorem 2.5 *Regarding K^i with constraints* (2.14) *and* (2.15), *the following statements hold.*

1. *If K^0 is designed to satisfy the equality constraints in* (2.14), *so will K^i be for all $i \geq 1$.*
2. *K^i will satisfy the inequality constraint in* (2.15) *for all $i \geq 1$ if*
   - *The pair (A, F) is detectable with $Q = F^T F$;*
   - *(A, H_0) is stabilizable with $X_0 = H_0 H_0^T$, where $H_0 \in \mathbb{R}^{n \times s}$ is a full rank matrix;*
   - *$(I_n - H_0(H_0^T H_0)^{-1} H_0^T)B = 0$.*

Proof of Theorem 2.5: During each iteration, D is guaranteed to satisfy (2.14), i.e. $\mathcal{C}(D^i) = 0$. Since the gain is updated as $K^i = K^{i-1} - \alpha^i D^i$, it is straightforward that if $\mathcal{C}(K^{i-1}) = 0$, then $\mathcal{C}(K^i) = 0$. Therefore, the updated gain matrix always satisfies the equality constraints in (2.14) as long as the initial gain matrix is designed to satisfy the equality constraints in (2.14). This proves Statement 1 of Theorem 2.5.

To prove that (2.15) always holds for any iteration $i \geq 1$, first, we need to prove that the finiteness of functional cost is equivalent to $\lim_{t\to\infty} e^{A_c t} x_0 = 0$. This is because if $x_0 \in \text{span}(H_0)$, $\lim_{t\to\infty} e^{A_c t} x_0 = 0$ is equivalent to $\lim_{t\to\infty} e^{A_c t} h_{oj} = 0$, where span($\cdot$) represents the span of a matrix, and h_{oj} is the jth column of H_0, $j = 1, \ldots, s$. It is also equivalent to $0 < J(K) = \text{Tr}(P(K)X_0) = \sum_{j=1}^{s} h_{oj}^T P h_{oj} < \infty$ due to the detectability of (A, F).

Second, we need to prove that $\lim_{t\to\infty} e^{A_c t} x_0 = 0$ guarantees the stability of the closed-loop system. It is true if either of the following two conditions is satisfied.

- $s = n$, i.e. $X_0 > 0$;
- (A_c, H_0) is stabilizable.

It is straightforward that the first condition guarantees the stability of the closed-loop system; for the second condition, $e^{A_c t} h_{oi}$ is the impulse response of the system $\dot{x}(t) = A_c x(t) + h_{oi}\delta(t)$ and all its modes are excited. Since $b_n \in \text{span}(H_0)$, where b_n is any column of B, then $\forall x \in \text{span}(B) \to x \in \text{span}(H_0)$, i.e. $\text{span}(B) \subseteq \text{span}(H_0)$. Since (A, B) is stabilizable, it holds for (A_c, B) too, so (A_c, H_0) is stabilizable for gain K^i generated for all $i \geq 1$. More details can be found in (Geromel, Bernussou, and Peres 1994).

As shown in Theorem 2.1, the functional cost is always decreasing during iterations. Thus, as long as the initial functional cost is finite, all the functional costs generated during iterations are finite. Therefore, the algorithm will generate stabilizing gains K^i for all $i \geq 1$. This proves Statement 2 of Theorem 2.5.

2.3 Case Study

For simplification, "(t)" is omitted in the expression of time-history signals in this section.

2.3.1 Statement of Problem

A DHG is a long-stroke, high-speed Cartesian robot system, which consists of two linear motors mounted on two parallel carriages, moving another orthogonal member simultaneously in tandem (Hu, Yao, and Wang 2010; Yuan, Chen, Yao, and Zhu 2017; Li, Yao, Zhu, and Wang 2015). Two carriages are linked by an orthogonal, motorized cross-arm with an end-effector laden payload. The moving gantry stage can be designed to provide high-speed, high-precision Cartesian motion to facilitate automated processes in surface mounting (Li and Yoon 2017), lithography (Teo, Yang, and Chen 2014), CNC machining (Farjood, Vojdani, Torabi, and Khaledi 2017), CT scanning (Kuriyama,

Onishi, Sano, Komiyama, Aikawa, Tateda, Araki, and Uematsu 2003) and X-ray stepping (Sullivan, Suiter, and Huston 2000), where high accuracy Cartesian motion is necessary (Yuan, Chen, Yao, and Zhu 2017). The DHG is able to carry a payload over a planar task space (Yao, Hu, and Wang 2012; Hu, Yao, and Wang 2011; Hu, Hu, Zhu, and Wang 2017).

Due to less computational intensity, ease of reconfigurability, and robustness towards a centralized failure (Bakule 2008), the decentralized control architecture is bridged in current industrial practice (Hou, Nian, Xiong, Wang, and Peng 2016; Pipelzadeh, Chaudhuri, and Green 2013; Yao, Jiang, Fang, Wen, and Cheng 2014), where the DHG can be modeled as a group of independent SISO systems, and the individual axis of motion is independently driven by a servo controller, via either the master-master scheme or the master-slave scheme (Giam, Tan, and Huang 2007). Sometimes, the coupling issues between the two carriages are not considered and the information regarding their interactions is not utilized in the controller design; then a conservative setting of controller parameters can be obtained, which possibly degrades the system performance (Silva and Erraz 2006).

In the literature, the DHG utilizes rigid joints to support its cross-arm with an end-effector actuator mounted on its cross-arm (Giam, Tan, and Huang 2007; Yao 2015). The rigid joints resist de-synchronization and the system is tolerant with high frequency control signals (Lu, Chen, Yao, and Wang 2013; Chen, Yao, and Wang 2015). However, the resulted inter-axis coupling force may damage the joints. Hence, an alternative design is to replace rigid joints with flexure joints (García-Herreros, Kestelyn, Gomand, Coleman, and Barre 2013). This allows a small rotary motion, which is prestigious for certain applications such as wafer scanning. The flexure-based mechanism has been widely applied in different platforms like electromagnetic (Zhu, Pang, and Teo 2017) or piezoelectric driven nanopositioning systems (Xiao and Li 2013). In our design, the pair of flexure joints enables a small degree of rotation of the cross-arm to prevent machine-wide damage due to unintended operations and allow for quick and cost-effective repair through joint replacement. Such design may introduce hidden modes to the setup, making the end effector sensitive to high frequency control signals (Chen, Yao, and Wang 2013). In some prior works, the effects of resonant modes are minimized using frequency domain based techniques, like gain stabilization or phase stabilization (Zhu, Pang, and Teo 2017; Al Mamun, Guo, and Bi 2006), or time domain based techniques, such as optimal control (Rădac, Precup, Petriu, Preitl, and Dragoş 2013; Rotondo, Nejjari, and Puig 2015; Li, Wu, He, and Chen 2012). Due to the above reasons, optimization of the control input chattering and the jerk of flexure force is essential. Notice that the effectiveness of the resonant mode suppression is not just determined by the stiffness of the flexure joints, but also the motion controller parameters with respect to given motion profiles.

Figure 2.1 shows the actual setup of the air bearing DHG being studied in this work and associated symbols for all parameters are shown in Table 2.1. The entire setup consists of three major components, which are the central

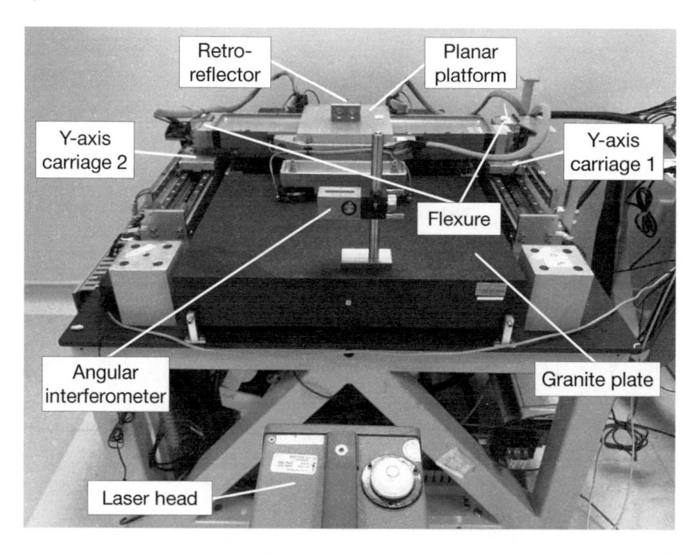

Figure 2.1: Setup of DHG.

structure, the two parallel carriages and the cross-arm with the end effector. Each of the three moving carriages is driven by an Akribis AUM3-S4 linear motor. Notably, the joints that connect the cross-arm to the carriages are made of small, flexible sheets of stainless steel. Seven aero-static bearings guide the motion and provide support on the end-effector platform so that the contact friction is minimized. The end-effector platform is located on flat granite workspace, hence there is very minimal height variation. This is particularly useful for camera-based inspection applications. Since the rotation yaw angle is usually small, the effect of flexure deformation to the X-axis motion is usually negligible. Hence, in the coming experiment, we only initiate our design based on the motion of the Y-axis, while holding the X-axis actuator at the center of the cross-arm. Each of the linear motors is connected to a Copley Accelus driver configured for torque mode, which receives its command from a dSPACE DS1005 modular programmable controller.

The actual schematic diagram of the DHG is shown in Figure 2.2(a), and it can be further simplified to a two-mass model linked by a pair of flexible joints, with a light rod for determining the end-effector position, as shown in Figure 2.2(b). Under this simplification, the DHG can be modeled as a coupled linear system as

$$M_1\ddot{y}_1 = K_\mathrm{f}u_1 - \Gamma_1\dot{y}_1 + v + d_1, \tag{2.52a}$$

$$M_2\ddot{y}_2 = K_\mathrm{f}u_2 - \Gamma_2\dot{y}_2 - v + d_2. \tag{2.52b}$$

with $M_1 = (m_\mathrm{e} + m_\mathrm{c})/2 + m_1$ and $M_2 = (m_\mathrm{e} + m_\mathrm{c})/2 + m_2$. Here, we assume that the damping coefficients of two axes are the same, where $\Gamma_1 = \Gamma_2 = \Gamma$. To the industrial standard, the control input u_1 and u_2 are combinations of

Table 2.1: Nomenclature used in Chapter 2 © [2018] IEEE. Reprinted, with permission, from J. Ma, S.-L. Chen, N. Kamaldin, C. S. Teo, A. Tay, A. Al. Mamun, and K. K. Tan, Integrated Mechatronic Design in the Flexure-Linked Dual-Drive Gantry by Constrained Linear-Quadratic Optimization, IEEE Transactions on Industrial Electronics, vol. 65, no. 3, pp. 2408–2418, 2018.

Name	Unit	Description
K_{f}	N/A	Force constant
l	m	Length of cross-arm
m_1	kg	Mass of carriage 1
m_2	kg	Mass of carriage 2
m_{e}	kg	Mass of end effector
m_{c}	kg	Mass of cross-arm
d_1	N	Disturbance in carriage 1
d_2	N	Disturbance in carriage 2
Γ_1	Ns/m	Damping coefficient in carriage 1
Γ_2	Ns/m	Damping coefficient in carriage 2
y_1	m	Position of carriage 1
y_2	m	Position of carriage 2
$y_{1\mathrm{d}}$	m	Reference position of carriage 1
$y_{2\mathrm{d}}$	m	Reference position of carriage 2
Y	m	Position of arm center
Θ	rad	Yaw angle between two carriages
Y_{d}	m	Reference position of arm center
Θ_{d}	rad	Reference yaw angle between two carriages
u_1	A	Control current of carriage 1
u_2	A	Control current of carriage 2
$u_{1\mathrm{ff}}$	A	Feedforward control current of carriage 1
$u_{2\mathrm{ff}}$	A	Feedforward control current of carriage 2
$u_{1\mathrm{fb}}$	A	Feedback control current of carriage 1
$u_{2\mathrm{fb}}$	A	Feedback control current of carriage 2
v	N	Coupling force from flexure
k_{v}	N/m	Stiffness of flexure

decentralized 2-DOF feedforward control and feedback control, that is

$$u_1 = u_{1\mathrm{ff}} + u_{1\mathrm{fb}}, \tag{2.53a}$$

$$u_2 = u_{2\mathrm{ff}} + u_{2\mathrm{fb}}. \tag{2.53b}$$

The end effector of the DHG is aimed to track given reference profiles, with minimizing the possible induced chattering from flexure force and control signals. These requirements are quantified as

- The center of the cross-arm $Y = (y_1 + y_2)/2$ and the yaw angle between the two carriages $\Theta = (y_1 - y_2)/l$ track two given S-curve reference trajectories Y_{d} and Θ_{d}, respectively.

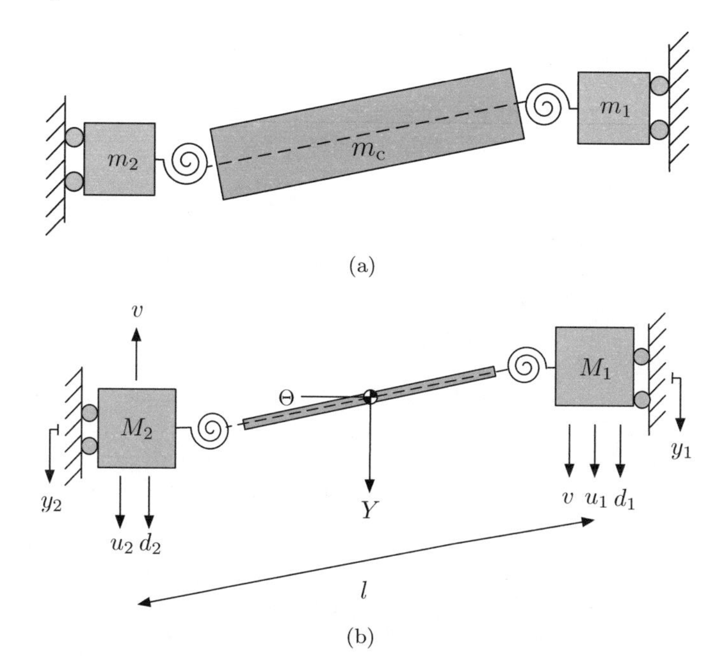

Figure 2.2: Schematic of DHG without X-axis carriage. (a) Actual model. (b) Simplified model.

- The chattering of feedback control signals $\dot{u}_{1\text{fb}}$, $\dot{u}_{2\text{fb}}$ and the jerk of the coupling force from the flexure $\dot{v}$ should be kept small to avoid the influences of hidden modes introduced by the flexure joints.

All the above factors are incorporated into the following cost function, where

$$J_1 = \int_0^\infty ((q_{11}\tilde{Y}^2 + q_{22}\tilde{\Theta}^2 + r_{11}\dot{u}_{1\text{fb}}^2 + r_{22}\dot{u}_{2\text{fb}}^2 + r_{33}\dot{v}^2)\, dt, \tag{2.54}$$

where $\tilde{Y} = Y_\text{d} - Y$ is the tracking error of the cross-arm center, and $\tilde{\Theta} = \Theta_\text{d} - \Theta$ is the tracking error of the yaw angle. Notice that the yaw angle is assumed to be within 5°.

Based on the known model, we design the feedforward controller

$$u_{1\text{ff}} = \frac{\Gamma_1}{K_\text{f}}\dot{y}_{1\text{d}} + \frac{M_1}{K_\text{f}}\ddot{y}_{1\text{d}}, \tag{2.55a}$$

$$u_{2\text{ff}} = \frac{\Gamma_2}{K_\text{f}}\dot{y}_{2\text{d}} + \frac{M_2}{K_\text{f}}\ddot{y}_{2\text{d}}, \tag{2.55b}$$

and the model is converted to

$$-M_1\ddot{e}_1 = K_{\mathrm{f}}u_{1\mathrm{fb}} + \Gamma_1\dot{e}_1 + v + d_1, \tag{2.56a}$$

$$-M_2\ddot{e}_2 = K_{\mathrm{f}}u_{2\mathrm{fb}} - \Gamma_2\dot{e}_2 - v + d_2, \tag{2.56b}$$

where $e_1 = y_{1\mathrm{d}} - y_1$ and $e_2 = y_{2\mathrm{d}} - y_2$, $y_{1\mathrm{d}}$ and $y_{2\mathrm{d}}$ are the joint-space reference profiles.

Assume Y_{d} and Θ_{d} are both smooth S-curves, moving from Y_{a} and Θ_a to their origins accordingly. That is $\dot{z}_{\mathrm{d}} = A_{\mathrm{d}}z_{\mathrm{d}}$, $\mathcal{Y}_{\mathrm{d}} = C_{\mathrm{d}}z_{\mathrm{d}}$, with $z_{\mathrm{d}}^T = \begin{bmatrix} z^T & \phi^T \end{bmatrix}$, $A_{\mathrm{d}} = \mathrm{blocdiag}\{A_{\mathrm{z}}, A_\phi\}$, $\mathcal{Y}_{\mathrm{d}}^T = \begin{bmatrix} Y_{\mathrm{d}} & \Theta_{\mathrm{d}} \end{bmatrix}$, $C_{\mathrm{d}} = \mathrm{blocdiag}\{C_{\mathrm{z}}, C_\phi\}$,
$A_{\mathrm{z}} = \begin{bmatrix} 0 & 1 & 0 \\ 0 & 0 & 1 \\ a_{\mathrm{z}1} & a_{\mathrm{z}2} & a_{\mathrm{z}3} \end{bmatrix}$, $A_\phi = \begin{bmatrix} 0 & 1 & 0 \\ 0 & 0 & 1 \\ a_{\phi 1} & a_{\phi 2} & a_{\phi 3} \end{bmatrix}$, $C_{\mathrm{z}} = C_\phi = \begin{bmatrix} 1 & 0 & 0 \end{bmatrix}$.

In this book, $\mathrm{blocdiag}\{A_1, \ldots, A_n\}$ represents a block diagonal matrix with matrices A_i, $\forall i = 1, \ldots, n$ as diagonal entries, and $\mathrm{diag}\{a_1, \ldots, a_n\}$ represents a diagonal matrix with numbers a_i, $\forall i = 1, \ldots, n$ as diagonal entries.

Remark 2.3 *For non-zero final end-effector positions Y_{b} and Θ_{b}, we can redefine $\bar{Y}_{\mathrm{d}} = Y_{\mathrm{d}} - Y_{\mathrm{b}}$, $\bar{Y} = Y - Y_{\mathrm{b}}$ and $\bar{\Theta}_{\mathrm{d}} = \Theta_{\mathrm{d}} - \Theta_{\mathrm{b}}$, $\bar{\Theta} = \Theta - \Theta_{\mathrm{b}}$, and use $\bar{(\cdot)}$ to replace their counterparts in the following discussion.*

In low-cost manufacturing, there is usually no real-time feedback of the end-effector position by laser sensors. In that case, the tracking error on the joint-space and task-space coordinates can be approximately related by $\tilde{Y} = (e_1 + e_2)/2$, $\tilde{\Theta} = (e_1 - e_2)/l$, and their reference trajectories are linked by $\begin{bmatrix} y_{1\mathrm{d}} & y_{2\mathrm{d}} \end{bmatrix}^T = T_1 \begin{bmatrix} Y_{\mathrm{d}} & \Theta_{\mathrm{d}} \end{bmatrix}^T$, $\begin{bmatrix} Y & \Theta \end{bmatrix}^T = T_2 \begin{bmatrix} y_1 & y_2 \end{bmatrix}^T$, where $T_1 = \begin{bmatrix} 1 & l/2 \\ 1 & -l/2 \end{bmatrix}$, $T_2 = \begin{bmatrix} 1/2 & 1/2 \\ 1/l & -1/l \end{bmatrix}$. However, such approximation is based on the assumption of no significant excitation of resonant modes. To this point, it is indeed essential to minimize the chattering of both control signals and flexure force in the DHG.

2.3.2 Formulation of Constrained Linear Quadratic Optimization Problem

The objective function in (2.54) can be expressed as

$$J_1 = \int_0^\infty (\tilde{\mathcal{Y}}^T Q_1 \tilde{\mathcal{Y}} + \dot{u}_{\mathrm{fb}}^T R \dot{u}_{\mathrm{fb}})\, dt, \tag{2.57}$$

where $\tilde{\mathcal{Y}} = \mathcal{Y}_{\mathrm{d}} - \mathcal{Y}$, with $\mathcal{Y}^T = \begin{bmatrix} Y & \Theta \end{bmatrix}$, $Q_1 = \mathrm{diag}\{q_{11}, q_{22}\} \geq 0$, $\dot{u}_{\mathrm{fb}}^T = \begin{bmatrix} \dot{u}_{1\mathrm{fb}} & \dot{u}_{2\mathrm{fb}} & \dot{v} \end{bmatrix}$, and $R = \mathrm{diag}\{r_{11}, r_{22}, r_{33}\} > 0$.

We convert the original system model to the following state-space form

$$\dot{x} = Ax + Bu + Ed, \tag{2.58a}$$

$$\mathcal{Y} = Cx, \tag{2.58b}$$

where $x^T = \begin{bmatrix} y_1 & \dot{y}_1 & y_2 & \dot{y}_2 \end{bmatrix}$, $u^T = \begin{bmatrix} u_1 & u_2 & v \end{bmatrix}$, $d^T = \begin{bmatrix} d_1 & d_2 \end{bmatrix}$. Matrices A, B, E and C are given by

$$A = \begin{bmatrix} 0 & 1 & 0 & 0 \\ 0 & -\frac{\Gamma_1}{M_1} & 0 & 0 \\ 0 & 0 & 0 & 1 \\ 0 & 0 & 0 & -\frac{\Gamma_2}{M_2} \end{bmatrix}, \quad B = \begin{bmatrix} 0 & 0 & 0 \\ \frac{K_f}{M_1} & 0 & \frac{1}{M_1} \\ 0 & 0 & 0 \\ 0 & \frac{K_f}{M_2} & -\frac{1}{M_2} \end{bmatrix},$$

$$E = \begin{bmatrix} 0 & 0 \\ \frac{1}{M_1} & 0 \\ 0 & 0 \\ 0 & \frac{1}{M_2} \end{bmatrix}, \quad C = \begin{bmatrix} \frac{1}{2} & 0 & \frac{1}{2} & 0 \\ \frac{1}{l} & 0 & -\frac{1}{l} & 0 \end{bmatrix}. \tag{2.59}$$

Under the assumption that the disturbances are slow-varying, we take the derivative of (2.58a) and augment y_1, y_2 as new state variables; then the new state vector is defined as

$$\begin{aligned} \hat{x}^T &= \begin{bmatrix} y_1 & \dot{y}_1 & \ddot{y}_1 & y_2 & \dot{y}_2 & \ddot{y}_2 \end{bmatrix} \\ &= \begin{bmatrix} x_1 & x_2 & x_3 & x_4 & x_5 & x_6 \end{bmatrix}. \end{aligned} \tag{2.60}$$

The equivalent state-space representation of the system is given by

$$\dot{\hat{x}} = \hat{A}\hat{x} + \hat{B}\dot{u}, \tag{2.61a}$$

$$\mathcal{Y} = \hat{C}\hat{x}, \tag{2.61b}$$

where matrices $\hat{A}$, $\hat{B}$ and $\hat{C}$ are given by

$$\hat{A} = \begin{bmatrix} 0 & 1 & 0 & 0 & 0 & 0 \\ 0 & 0 & 1 & 0 & 0 & 0 \\ 0 & 0 & -\frac{\Gamma_1}{M_1} & 0 & 0 & 0 \\ 0 & 0 & 0 & 0 & 1 & 0 \\ 0 & 0 & 0 & 0 & 0 & 1 \\ 0 & 0 & 0 & 0 & 0 & -\frac{\Gamma_2}{M_2} \end{bmatrix},$$

$$\hat{B} = \begin{bmatrix} 0 & 0 & 0 \\ 0 & 0 & 0 \\ \frac{K_f}{M_1} & 0 & \frac{1}{M_1} \\ 0 & 0 & 0 \\ 0 & 0 & 0 \\ 0 & \frac{K_f}{M_2} & -\frac{1}{M_2} \end{bmatrix}, \quad \hat{C} = \begin{bmatrix} \frac{1}{2} & 0 & 0 & \frac{1}{2} & 0 & 0 \\ \frac{1}{l} & 0 & 0 & -\frac{1}{l} & 0 & 0 \end{bmatrix}. \tag{2.62}$$

Define $\bar{C} = I_6 - L\hat{C}$, where $L = \hat{C}^T(\hat{C}\hat{C}^T)^{-1}$. In other words, $\bar{C}$ is the projection matrix onto the orthogonal complement of range space of $\hat{C}^T$. Then $\bar{\mathcal{Y}} = \bar{C}\hat{x}$, where $\bar{\mathcal{Y}}$ is the part of the state vector that is not seen by $\mathcal{Y} = \hat{C}\hat{x}$. The cost function is re-formulated by adding in the quadratic term of $\bar{\mathcal{Y}}$ into (2.57) as

$$J_2 = \int_0^\infty (\tilde{\mathcal{Y}}^T Q_1 \tilde{\mathcal{Y}} + \bar{\mathcal{Y}}^T Q_2 \bar{\mathcal{Y}} + \dot{u}_{\text{fb}}^T R \dot{u}_{\text{fb}})\, dt, \tag{2.63}$$

with $Q_2 = \text{diag}\{q_{2,ii}\} \geq 0$, $\forall i = 1,\ldots,6$. Notice that the additional term $\bar{y}^T Q_2 \bar{y}$ aims to convert the output feedback to the state feedback, the generalization in (2.63) can be made without loss of generality since we can always set Q_2 to be sufficiently small.

J_2 is converted to a tracking optimization problem based on joint-space state feedback as

$$J_3 = \int_0^\infty (\tilde{x}^T Q \tilde{x} + \dot{u}_{\text{fb}}^T R \dot{u}_{\text{fb}})\, dt, \tag{2.64}$$

with $\tilde{x} = \hat{x}_{\text{d}} - \hat{x}$, $\hat{x}_{\text{d}}^T = \begin{bmatrix} y_{1\text{d}} & \dot{y}_{1\text{d}} & \ddot{y}_{1\text{d}} & y_{2\text{d}} & \dot{y}_{2\text{d}} & \ddot{y}_{2\text{d}} \end{bmatrix}$ and $Q = \hat{C}^T Q_1 \hat{C} + \bar{C}^T Q_2 \bar{C} \geq 0$.

Define $e^T = \begin{bmatrix} e_1 & \dot{e}_1 & e_2 & \dot{e}_2 \end{bmatrix}$; the error dynamic model (2.56) is written as $\dot{e} = Ae - Bu_{\text{fb}} - Ed$. So we have $\dot{\hat{e}} = \hat{A}\hat{e} - \hat{B}\dot{u}_{\text{fb}}$, where $\hat{e}^T = \begin{bmatrix} e_1 & \dot{e}_1 & \ddot{e}_1 & e_2 & \dot{e}_2 & \ddot{e}_2 \end{bmatrix}$.

The task-space reference trajectory can be converted to joint-space reference trajectory $\dot{\hat{x}}_{\text{d}} = \hat{A}_{\text{d}} \hat{x}_{\text{d}}$. In the case that $A_{\text{z}} = A_\phi$, we have $\hat{A}_{\text{d}} = A_{\text{d}}$. Eventually, the error dynamic model is augmented with joint-space reference trajectory as $\dot{\bar{x}} = \bar{A}\bar{x} + \bar{B}\dot{u}$, where $\bar{x}^T = \begin{bmatrix} \hat{e}^T & \hat{x}_{\text{d}}^T \end{bmatrix}$, $\bar{A} = \text{blocdiag}\{\hat{A}, \hat{A}_{\text{d}}\}$, $\bar{B}^T = \begin{bmatrix} -\hat{B}^T & 0_{3\times 6} \end{bmatrix}$. Thus, J_3 is converted to the standard LQR format, where

$$J_4 = \int_0^\infty (\bar{x}^T \bar{Q} \bar{x} + \dot{u}_{\text{fb}}^T R \dot{u}_{\text{fb}})\, dt, \tag{2.65}$$

and $\bar{Q} = \text{blocdiag}\{Q, Q_{\text{d}}\} \geq 0$. Since the convergence of reference trajectory is not our objective, Q_{d} is set to be sufficiently small.

From $\dot{u}_{\text{fb}} = K\bar{x}$, we can restore it to the standard feedback controller structure, which is given by

$$u_{\text{fb}} = K \int_0^t \bar{x}\, d\tau. \tag{2.66}$$

Notice that v is treated as a "mechanical" control input in the problem formulation. But in the actual system, v is only regarded as part of system dynamics, which is not used for control purposes. The feedback control K_{FB} is block diagonal because we need to make sure the controllers $u_{1\text{fb}}$ and $u_{2\text{fb}}$ are decentralized in nature. In addition, the flexure joint results in coupling force

$$v = k_{\text{v}}(y_2 - y_1) = k_{\text{v}}(y_{2\text{d}} - y_{1\text{d}} - e_2 + e_1). \tag{2.67}$$

Thus, the controller gain K is structured as

$$K = \begin{bmatrix} K_{\text{FB}} & 0_{2\times 6} \\ K_{\text{V}} & \end{bmatrix}, \tag{2.68}$$

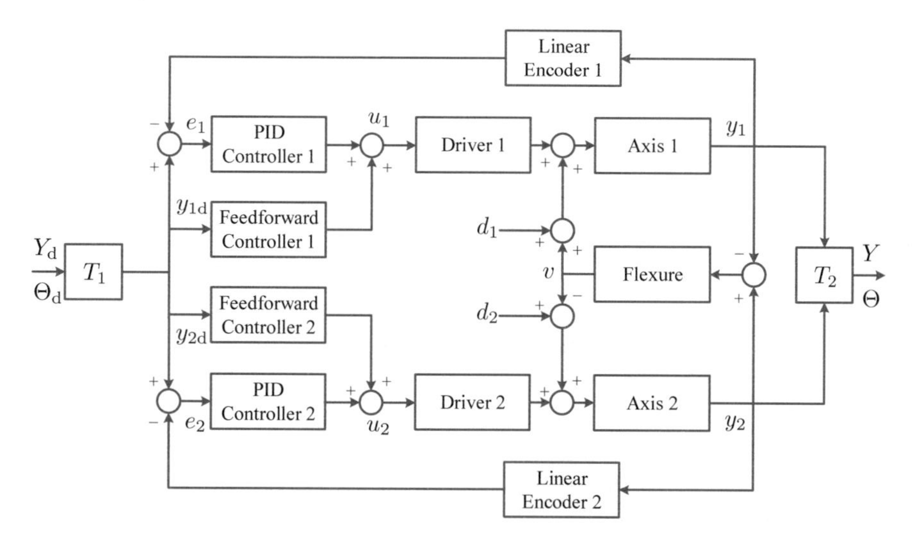

Figure 2.3: Overview of decentralized control architecture.

with

$$K_{\mathrm{FB}} = \mathrm{blocdiag}\{ \begin{bmatrix} k_{\mathrm{i}1} & k_{\mathrm{p}1} & k_{\mathrm{d}1} \end{bmatrix}, \begin{bmatrix} k_{\mathrm{i}2} & k_{\mathrm{p}2} & k_{\mathrm{d}2} \end{bmatrix} \}, \tag{2.69}$$

$$K_{\mathrm{V}} = \begin{bmatrix} 0 & k_{\mathrm{v}} & 0 & 0 & -k_{\mathrm{v}} & 0 & 0 & -k_{\mathrm{v}} & 0 & 0 & k_{\mathrm{v}} & 0 \end{bmatrix}. \tag{2.70}$$

In other words, the gain matrix K must be restricted with certain structural constraints, including the zero elements in certain positions of the gain matrix as well as the equality constraints due to the use of flexure. To this point, the block diagram of the overall closed-loop system for this DHG is shown in Figure 2.3.

To initiate the parameter optimization, the structural constraint on the gain matrix K is written as $K \in \Phi$, where $\Phi = \{K \in \mathbb{R}^{3\times 12} : \mathcal{C}(K) = \mathcal{C}_0\}$, and $\mathcal{C}(K)$ is a function of K with $\mathcal{C}(0) = 0$. The constraint of zero elements in K is expressed as

$$\mathcal{C}_1(K) = \tilde{A}_1 K \tilde{B}_1 + \tilde{A}_2 K \tilde{B}_2 + \tilde{A}_3 K \tilde{B}_3 = 0, \tag{2.71}$$

where

$$\begin{aligned}
\tilde{A}_1 &= \mathrm{diag}\{1,\ 0,\ 0\},\\
\tilde{A}_2 &= \mathrm{diag}\{0,\ 1,\ 0\},\\
\tilde{A}_3 &= \mathrm{diag}\{0,\ 0,\ 1\},\\
\tilde{B}_1 &= \mathrm{diag}\{0,\ 0,\ 0,\ 1,\ 1,\ 1,\ 1,\ 1,\ 1,\ 1,\ 1,\ 1\},\\
\tilde{B}_2 &= \mathrm{diag}\{1,\ 1,\ 1,\ 0,\ 0,\ 0,\ 1,\ 1,\ 1,\ 1,\ 1,\ 1\},\\
\tilde{B}_3 &= \mathrm{diag}\{1,\ 0,\ 1,\ 1,\ 0,\ 1,\ 1,\ 0,\ 1,\ 1,\ 0,\ 1\}.
\end{aligned} \tag{2.72}$$

The equality constraints due to the use of flexure reflected in the 3rd row of K are expressed as

$$\mathcal{C}_2(K) = \tilde{A}_4 K \tilde{B}_4 + \tilde{A}_5 K \tilde{B}_5 = 0, \tag{2.73a}$$

$$\mathcal{C}_3(K) = \tilde{A}_6 K \tilde{B}_6 + \tilde{A}_7 K \tilde{B}_7 = 0, \tag{2.73b}$$

$$\mathcal{C}_4(K) = \tilde{A}_8 K \tilde{B}_8 + \tilde{A}_9 K \tilde{B}_9 = 0, \tag{2.73c}$$

where

$$\begin{aligned}
\tilde{A}_4 = \tilde{A}_5 = \tilde{A}_6 = \tilde{A}_7 = \tilde{A}_8 = \tilde{A}_9 &= \begin{bmatrix} 0 & 0 & 1 \end{bmatrix}, \\
\tilde{B}_4 = \tilde{B}_6 &= \begin{bmatrix} 0 & 1 & 0 & 0 & 0 & 0 & 0 & 0 & 0 & 0 & 0 & 0 \end{bmatrix}^T, \\
\tilde{B}_5 = \tilde{B}_8 &= \begin{bmatrix} 0 & 0 & 0 & 0 & 1 & 0 & 0 & 0 & 0 & 0 & 0 & 0 \end{bmatrix}^T, \\
\tilde{B}_7 &= \begin{bmatrix} 0 & 0 & 0 & 0 & 0 & 0 & 0 & 1 & 0 & 0 & 0 & 0 \end{bmatrix}^T, \\
\tilde{B}_9 &= \begin{bmatrix} 0 & 0 & 0 & 0 & 0 & 0 & 0 & 0 & 0 & 0 & 1 & 0 \end{bmatrix}^T.
\end{aligned} \tag{2.74}$$

Besides the equality constraints (2.71) and (2.73), the following inequality constraint (2.75) is also required to be guaranteed to ensure the closed-loop stability of the system.

$$\mathrm{Re}(\mathrm{eig}(A_{\mathrm{c}})) < 0, \tag{2.75}$$

with $A_{\mathrm{c}} = \bar{A} + \bar{B}K$.

The cost function (2.65) is equivalent to

$$J(K) = \mathrm{Tr}(P(K)X_0), \tag{2.76}$$

where $P(K) = \int_0^\infty \Lambda^T Q_{\mathrm{c}} \Lambda \, dt$, $X_0 = x_0 x_0^T$, $\Lambda = e^{A_{\mathrm{c}} t}$, $Q_{\mathrm{c}} = \bar{Q} + K^T R K$, x_0 is the vector corresponding to the initial value of the state $\bar{x}$. Hence, the optimization problem targets at minimizing $J(K)$ in (2.76) under constraints (2.71), (2.73) and (2.75).

This optimization problem is equivalent to

$$\min_D \frac{1}{2} \left\| \frac{dJ(K)}{dK} - D \right\|^2, \quad s.t. \, \mathcal{C}_i(D) = 0, \quad \forall i = 1, \ldots, 4. \tag{2.77}$$

Problem (2.77) is further converted to its equivalent dual problem as

$$\max_{\Lambda_i} \min_D \left(\frac{1}{2} \left\| \frac{dJ(K)}{dK} - D \right\|^2 + \mathrm{Tr}\left(\sum_{i=1}^{4} \Lambda_i^T \mathcal{C}_i(D) \right) \right), \tag{2.78}$$

where Λ_i is the dual variable associated with the equality constraint $\mathcal{C}_i(D) = 0$.

Notice that to execute the proposed algorithm, we need to project $dJ(K)/dK$ onto $\mathcal{C}_i(D) = 0$ to get D by Theorem 2.4, specifically, it is given by Corollary 2.1.

Corollary 2.1 *The projection gradient matrix is given by*

$$D = \frac{dJ(K)}{dK} - \sum_{j=1}^{3} \tilde{A}_j \frac{dJ(K)}{dK} \tilde{B}_j - \sum_{j=4}^{5} \tilde{A}_j^T \Lambda_2 \tilde{B}_j^T - \sum_{j=6}^{7} \tilde{A}_j^T \Lambda_3 \tilde{B}_j^T - \sum_{j=8}^{9} \tilde{A}_j^T \Lambda_4 \tilde{B}_j^T, \tag{2.79}$$

and dual variable is given by

$$\Lambda_i = \frac{1}{2} \left(\tilde{A}_{2i} \frac{dJ(K)}{dK} \tilde{B}_{2i} + \tilde{A}_{2i+1} \frac{dJ(K)}{dK} \tilde{B}_{2i+1} \right), \tag{2.80}$$

where $i = 2, \ldots, 4$.

Proof of Corollary 2.1: The necessary and sufficient optimal solution to the dual problem (2.78) is given by

$$D - \frac{dJ(K)}{dK} + \frac{\partial}{\partial D} \left(\mathrm{Tr} \left(\sum_{i=1}^{4} \Lambda_i^T c_i(D) \right) \right) = 0. \tag{2.81}$$

Thus,

$$D - \frac{dJ(K)}{dK} + \sum_{j=1}^{3} A_j^T \Lambda_1 B_j^T + \sum_{j=4}^{5} A_j^T \Lambda_2 B_j^T + \sum_{j=6}^{7} A_j^T \Lambda_3 B_j^T + \sum_{j=8}^{9} A_j^T \Lambda_4 B_j^T = 0. \tag{2.82}$$

By $A_1 \times (2.82) \times B_1 + A_2 \times (2.82) \times B_2 + A_3 \times (2.82) \times B_3$,

$$\sum_{j=1}^{3} A_j^T \Lambda_1 B_j^T = \sum_{j=1}^{3} A_j \frac{dJ(K)}{dK} B_j. \tag{2.83}$$

Substitute (2.83) into (2.82), and it gives (2.79). This proves Corollary 2.1.

2.3.3 Optimization and Experimental Validation

The above method is applicable to the situation when we are able to fabricate arbitrary flexures with different stiffness. In our experiments, only three pieces of flexures with different thickness (2 mm, 3 mm and 4 mm) are available for test. This is quite common in practice when the available flexures are only with the standard values of stiffness, differing from the value obtained from optimization. In this case, our objective is to find the most suitable one among them and determine the optimal controller parameters to minimize the cost function (2.54). Hence, we propose the following experimental procedures.

Given S-curve motion profiles, we first directly fix flexure stiffness to the values of those three pieces and optimize the controllers. Second, for every piece of flexure, we do the optimization to obtain controller parameters and compare the costs to see which flexure gives the lowest one. Third, we run the profiles using the optimized parameters. Last but not least, we check the yaw angular velocity measured by the laser interferometer. If it is acceptable, we complete the optimization. If not, we should adjust the weighting matrices and re-initialize the optimization and the experiment.

Thus, we need to set $\Lambda_2 = \Lambda_3 = \Lambda_4 = 0$ in (2.79) and change B_3 to B_{3+}, which is defined as $B_{3+} = \text{diag}\{1, 1, 1, 1, 1, 1, 1, 1, 1, 1, 1, 1\}$. A number of 5000 iterations is set as a stopping criterion for gain matrix update.

To effectively identify the stiffness k_v, we drive the two carriages to follow two low frequency sinusoidal signals with the same amplitude but 180° phase difference. The carriages are under motor forces, friction forces as well as the force from the flexure. To treat it as a whole system, we have $K_\text{f}(u_1 - u_2) = k_\text{v}(y_1 - y_2)$. Notice that u_1, u_2, y_1 and y_2 can be read from input and encoder feedback, and K_f is given in the datasheet, where $K_\text{f} = 62.8$ N/A, so we can do a least square curve fitting to calculate the stiffness k_v. The results show that flexures with widths of 2 mm, 3 mm and 4 mm have stiffness of 4887.3 N/m, 8693.7 N/m and 13349.3 N/m, respectively.

To identify the system dynamics, from (2.52), we have $M_1\ddot{y}_1 + M_2\ddot{y}_2 + \Gamma(\dot{y}_1 + \dot{y}_2) = K_\text{f}(u_1 + u_2) + d_1 + d_2$; then we can write it in the regression form as $\begin{bmatrix}\ddot{y}_1 & \ddot{y}_2 & \dot{y}_1 + \dot{y}_2\end{bmatrix}\begin{bmatrix}\hat{M}_1 & \hat{M}_2 & \hat{\Gamma}\end{bmatrix}^T = K_\text{f}(u_1 + u_2) + d_1 + d_2$. The same pseudo-random binary signal is injected to u_1 and u_2 for system identification. By assuming $(d_1 + d_2)$ is a zero-mean wide-sense stationary signal being independent of u_1 and u_2, we can use least square fitting via ARMAX method to get $\hat{M}_1 = 16.5$ kg, $\hat{M}_2 = 18.3$ kg, $\hat{\Gamma} = 172.7$ Ns/m.

From the generation of motion profiles, the eigenvalues of both matrices A_z and A_ϕ are all set to -2, with position steps of 0.1 m and 0.01 rad, respectively. The desired profiles for displacement and rotation are shown in Figure 2.4.

Weighting matrices can be chosen based on different requirements. The weighting matrices set 1 is chosen as $Q_1 = \text{diag}\{2 \times 10^7, 10^7\}$, $R = \text{diag}\{1, 1, 1\}$, $Q_2 = Q_\text{d} = 10^{-7}I_6$. The initial feedback controller in the proposed optimization is defined as

$$K_\text{FB}^0 = \text{blocdiag}\{\begin{bmatrix}40 & 1000 & 10\end{bmatrix}, \begin{bmatrix}40 & 1000 & 10\end{bmatrix}\}. \tag{2.84}$$

By using the 2-mm, the 3-mm and the 4-mm flexure joints, the initial gain matrices stabilize the closed-loop system with functional costs of 2.51×10^6, 2.51×10^6 and 2.51×10^6, respectively. By executing the optimization algorithm, after 5000 iterations, their functional costs are decreased to 2.96×10^4, 3.67×10^4, 4.27×10^4, respectively. Thus, we choose the 2-mm flexure for motion control experiment. By using this set of weighting matrices, the gain

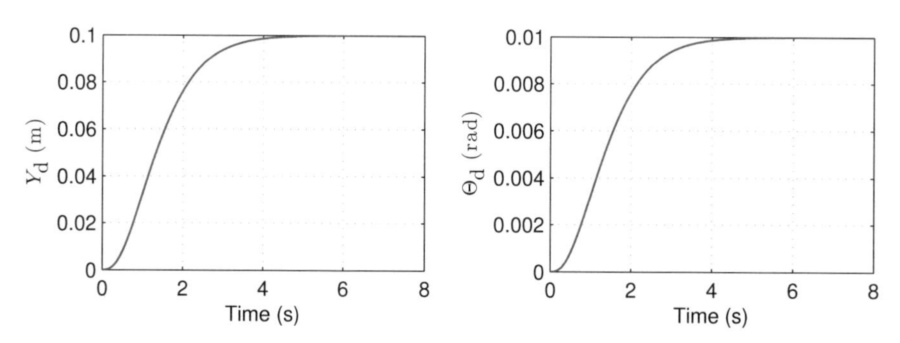

Figure 2.4: Reference profiles for cross-arm center position and yaw angle. © [2018] IEEE. Reprinted, with permission, from J. Ma, S.-L. Chen, N. Kamaldin, C. S. Teo, A. Tay, A. Al. Mamun, and K. K. Tan, Integrated Mechatronic Design in the Flexure-Linked Dual-Drive Gantry by Constrained Linear-Quadratic Optimization, IEEE Transactions on Industrial Electronics, vol. 65, no. 3, pp. 2408–2418, 2018.

is updated as

$$K_{\text{FB}}^{5000} = \text{blocdiag}\{ \begin{bmatrix}1628 & 331 & 12\end{bmatrix}, \begin{bmatrix}1503 & 282 & 10\end{bmatrix} \}. \tag{2.85}$$

Weighting matrices set 2 is chosen to penalize more on the tracking performance of the cross-arm center position. Thus, we change $Q_1 = \text{diag}\{5 \times 10^7, 10^7\}$ with all the other weighting matrices being preserved. By starting from the same initial condition, the initial function costs are given as 6.27×10^6, 6.27×10^6 and 6.27×10^6 by using the 2-mm, the 3-mm, and the 4-mm flexure joints, respectively. After 5000 iterations of optimization, their functional costs are decreased to 6.04×10^4, 7.27×10^4, 8.47×10^4, respectively. Thus, in this case, we choose the 2-mm flexure again to redo the experiment. With this choice, the gain with weighting matrices set 2 is updated as

$$K_{\text{FB}}^{5000} = \text{blocdiag}\{ \begin{bmatrix}1974 & 344 & 9\end{bmatrix}, \begin{bmatrix}1841 & 303 & 8\end{bmatrix} \}. \tag{2.86}$$

Figure 2.5 shows the functional cost J and the norm of the projection gradient matrix $\|D\|$ using three flexures during the first 5 iterations based on weighting matrices set 1 and set 2. It can be seen that the functional costs are monotonically decreasing with iterations, which is tallied with Theorem 2.1.

Figure 2.6 shows the tracking errors of the cross-arm center position and the yaw angle between the two carriages. The proposed algorithm indeed generates the stabilizing gains tallied with Theorem 2.5. The feedback control efforts and their chattering $\dot{u}_{1\text{fb}}$ and $\dot{u}_{2\text{fb}}$ are illustrated in Figure 2.7. Notice that $\dot{u}_{1\text{fb}}$ and $\dot{u}_{2\text{fb}}$ are obtained from the differentiation of $u_{1\text{fb}}$ and $u_{2\text{fb}}$ without the use of any low-pass filter. The coupling force v from the flexure calculated from (2.67) and its chattering are also shown in Figure 2.7, where the chattering is obtained from the differentiation of v without the use of any filter as well.

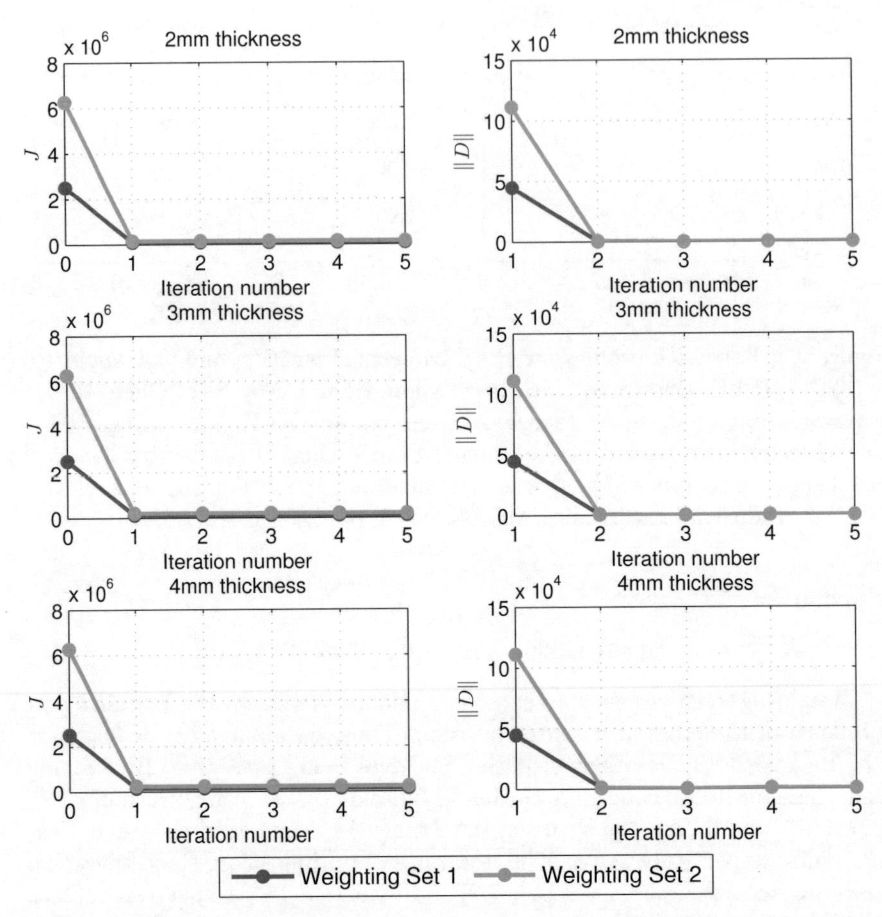

Figure 2.5: Functional cost and norm of projection gradient matrix during iterations. © [2018] IEEE. Reprinted, with permission, from J. Ma, S.-L. Chen, N. Kamaldin, C. S. Teo, A. Tay, A. Al. Mamun, and K. K. Tan, Integrated Mechatronic Design in the Flexure-Linked Dual-Drive Gantry by Constrained Linear-Quadratic Optimization, IEEE Transactions on Industrial Electronics, vol. 65, no. 3, pp. 2408–2418, 2018.

Compared with weighting matrices set 1, weighting matrices set 2 penalizes more on the tracking error of the cross-arm center position, and this matches the experimental results as shown in Figure 2.6. Figure 2.8 shows the angular velocities measured by the laser interferometer. Notice that the optimized results are meaningful as long as the chattering is within the range to avoid excitation of resonant modes. There are trade-offs between the different factors in the objective function. Good tracking performance of the cross-arm center position and the yaw angle in the joint space will sacrifice the performance of the chattering of control inputs and the jerk of flexure force resulting in the

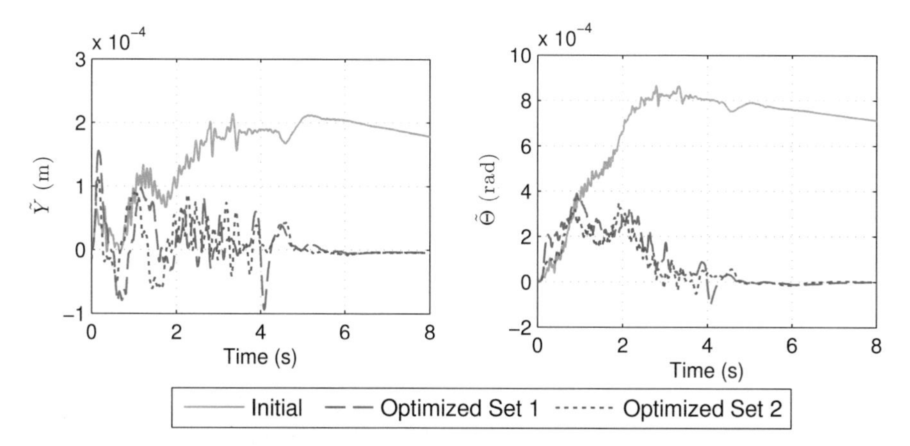

Figure 2.6: Comparison of tracking errors in task space. © [2018] IEEE. Reprinted, with permission, from J. Ma, S.-L. Chen, N. Kamaldin, C. S. Teo, A. Tay, A. Al. Mamun, and K. K. Tan, Integrated Mechatronic Design in the Flexure-Linked Dual-Drive Gantry by Constrained Linear-Quadratic Optimization, IEEE Transactions on Industrial Electronics, vol. 65, no. 3, pp. 2408–2418, 2018.

chattering of the end effector in the task space, and vice versa. In general, weighting matrices can be adjusted by the user for different penalties based on different optimization targets, and optimized parameters by the proposed algorithm will give a better overall performance according to the predefined cost function.

2.3.4 Summary

In this work, the flexure joint is mounted on the DHG to allow a small degree of angle rotation. To effectively minimize possible induced chattering to the end-effector plane, while maintaining good tracking accuracy to given motion profiles, the stiffness of the flexure and the feedback controller parameters need to be optimized simultaneously. This mechatronic design problem is converted to a linear quadratic optimization problem, and the objective function is initiated with consideration of the tracking errors, the chattering of feedback control signals and the jerk of coupling force resulting from the flexure joint. Since there is no closed-form solution to this constrained linear quadratic optimization problem, it is solved by the proposed constrained linear quadratic optimization algorithm. As validated by comparative experiments, optimized parameters by the proposed algorithm are capable of giving good performance according to the predefined cost function. The proposed method provides a possible solution to achieve better tracking and chattering suppression on the precision stage without costly system redesign.

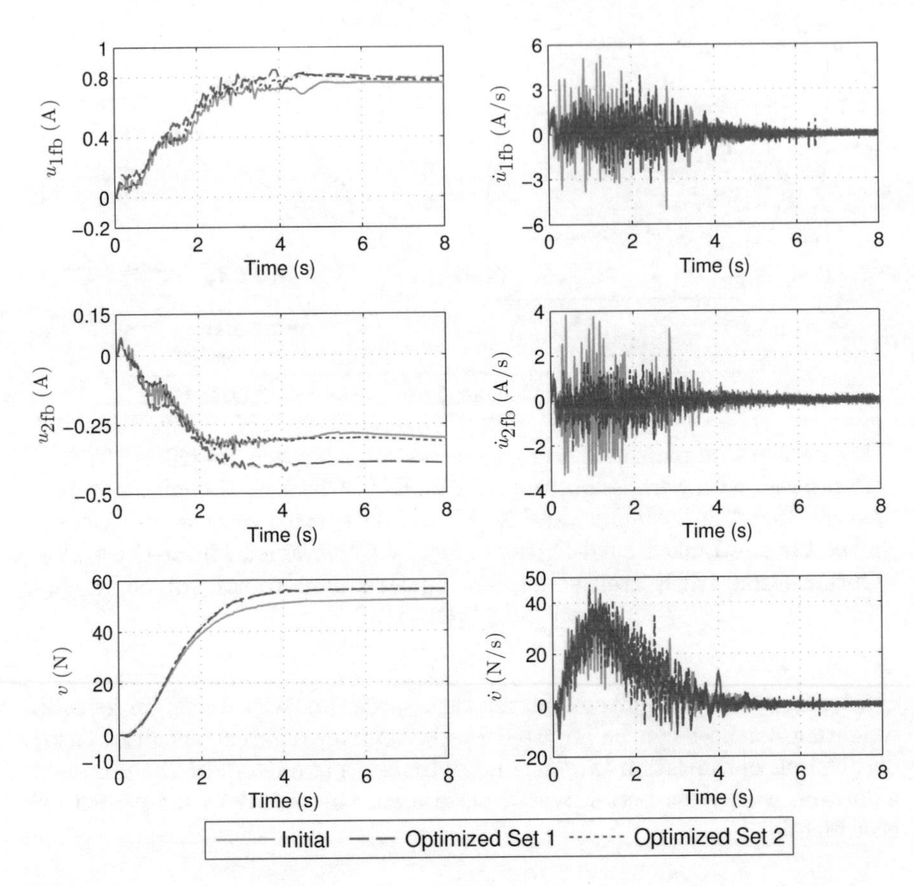

Figure 2.7: Comparison of control efforts, flexure force and their chattering. © [2018] IEEE. Reprinted, with permission, from J. Ma, S.-L. Chen, N. Kamaldin, C. S. Teo, A. Tay, A. Al. Mamun, and K. K. Tan, Integrated Mechatronic Design in the Flexure-Linked Dual-Drive Gantry by Constrained Linear-Quadratic Optimization, IEEE Transactions on Industrial Electronics, vol. 65, no. 3, pp. 2408–2418, 2018.

2.4 Conclusion

This chapter focuses on a linear quadratic optimization problem with a series of constraints imposed on the composite gain matrix. Theorems are given to calculate the gradient of the cost function with respect to the gain matrix for the standard LQR problem in both finite and infinite horizons. To cater to the constraints in the composite gain matrix, a projection method is introduced, and the projection of the gradient matrix onto the constrained hyperplane is

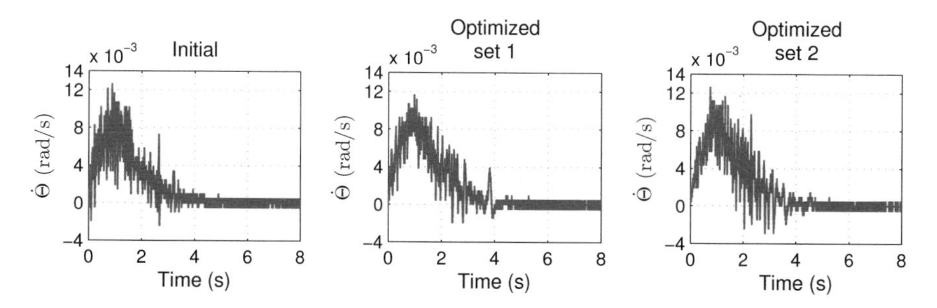

Figure 2.8: Comparison of yaw angular velocities measured by laser interferometer. © [2018] IEEE. Reprinted, with permission, from J. Ma, S.-L. Chen, N. Kamaldin, C. S. Teo, A. Tay, A. Al. Mamun, and K. K. Tan, Integrated Mechatronic Design in the Flexure-Linked Dual-Drive Gantry by Constrained Linear-Quadratic Optimization, IEEE Transactions on Industrial Electronics, vol. 65, no. 3, pp. 2408–2418, 2018.

summarized by another theorem. The original optimization problem is converted to the approximated equivalent counterpart, such that the distance between the projection gradient onto the constrained hyperplane to its unconstrained counterpart is minimized. Then, a gradient-based optimization algorithm is presented based on the direct computation of the projection gradient matrix and line search of the optimal step size. This algorithm starts from an initial solution satisfying both the equality constraint and the inequality constraint, and it determines a feasible direction of progress, from which it generates all the gain matrices that satisfy all the constraints with decreasing objective function. The proposed constrained linear quadratic optimization algorithm is applied to integrally design the flexure joints and the controllers in a DHG and the effectiveness of the systematic design approach is successfully validated by comparative experiments.

3

Constrained $\mathcal{H}_2$ Optimization

3.1 Background

Robust control laws aim to keep the error and the output signals of the system under pre-specified tolerance levels, despite the effect of uncertainties. These uncertainties can be external (disturbances and noises) or internal (plant model imperfections) to the system. Most of the robust control techniques are based on $\mathcal{H}_2$, $\mathcal{H}_\infty$, $\mathcal{H}_2/\mathcal{H}_\infty$ and μ-synthesis approaches.

For an $\mathcal{H}_2$ problem, the performance of a feedback system can be quantified in terms of the closed-loop gain from the disturbance inputs to the controlled outputs. The system $\mathcal{H}_2$-norm represents an average gain and it can be used as a performance function for an optimal control problem. The optimal control for the LQR is also optimal in terms of minimizing the $\mathcal{H}_2$-norm of the closed-loop system when the plant is appropriately defined. Formulating the LQR problem as an $\mathcal{H}_2$-norm optimization problem adds insight into the LQR, and yields a formulation that can be easily generalized to include frequency domain performance specifications.

Consider the stabilizable and detectable LTI system

$$\dot{x}(t) = Ax(t) + B_2u(t) + B_1w(t), \tag{3.1a}$$
$$z(t) = C_1x(t) + D_1u(t), \tag{3.1b}$$
$$y(t) = C_2x(t) + D_2w(t), \tag{3.1c}$$

with the control law

$$u(t) = Kx(t), \tag{3.2}$$

where x is the state variable with $x(0) = x_0$, y is the measured output, z is the controlled output, u is the control input and w is the disturbance.

The objective of an $\mathcal{H}_2$ control problem is to design a proper control law (3.2) such that the resulting closed-loop system is stable, and the $\mathcal{H}_2$-norm of the closed-loop transfer function $G(s)$ from the disturbance w to the controlled output z is minimized, given a class of disturbance inputs.

The $\mathcal{H}_2$-norm can be interpreted in two ways as illustrated below.

- $\mathcal{H}_2$ optimal control against deterministic disturbance inputs: Consider a stable causal LTI system with transfer function $G(s)$ from the disturbance

w to the controlled output z and its impulse response $G(t)$, the $\mathscr{H}_2$-norm of $G(s)$ measures the energy of the impulse response.

- $\mathscr{H}_2$ optimal control against stochastic disturbance inputs: Consider a hypothetical stochastic disturbance input signal w such that $E[w(t)] = 0$ and $E[w(t)w(t+\tau)^T] = I\delta(\tau)$; this is called white noise. Notice that it is a signal with infinite energy and finite power. The $\mathscr{H}_2$-norm of $G(s)$ measures the expected power of the response to white noise.

The cost function is interpreted as the square of the $\mathscr{H}_2$-norm of $G(s)$, which is given by

$$\begin{aligned} \|G(s)\|_2^2 &= \mathrm{Tr}\left(\int_0^\infty B_1^T e^{A^T t} C_1^T C_1 e^{At} B_1 \, dt \right) \\ &= \mathrm{Tr}(B_1^T W_o B_1), \end{aligned} \tag{3.3}$$

where W_o is the observability Gramian satisfying

$$A^T W_o + W_o A + C_1^T C_1 = 0. \tag{3.4}$$

The cost function can also be interpreted by controllability Gramian W_c, where

$$\begin{aligned} \|G(s)\|_2^2 &= \mathrm{Tr}\left(\int_0^\infty C_1 e^{At} B_1 B_1^T e^{A^T t} C_1^T \, dt \right) \\ &= \mathrm{Tr}(C_1 W_c C_1^T), \end{aligned} \tag{3.5}$$

with

$$A W_c + W_c A^T + B_1 B_1^T = 0. \tag{3.6}$$

To obtain the general solution of an $\mathscr{H}_2$ state feedback problem, it is assumed that the pair (A, B_2) is stabilizable, $D_1^T D_1$ is non-singular, $C_1 D_1^T = 0$, the pair (A, C_1) has no unobservable modes on the imaginary axis.

The optimal static state feedback controller gain is given by

$$K = -(D_1^T D_1)^{-1} B_2^T P, \tag{3.7}$$

and P is the unique symmetric positive semi-definite solution to the ARE

$$A^T P + PA - PB_2(D_1^T D_1)^{-1} B_2^T P + C_1^T C_1 = 0. \tag{3.8}$$

The optimal cost is given by

$$J^* = \mathrm{Tr}(B_1^T P B_1). \tag{3.9}$$

Notice that the LQR problem is the special case of $\mathscr{H}_2$ synthesis under the assumption of full state feedback, i.e. $C_2 = I$, and no disturbance input, i.e. $w(t) = 0$. Consider the objective is to find a control signal $u(t)$ that minimizes

$$\|G(s)\|_2^2 = \int_0^\infty \|C_1 x(t) + D_1 u(t)\|_2^2 \, dt, \tag{3.10}$$

given the initial condition $x(0)$. If $C_1 = \begin{bmatrix}\sqrt{Q} & 0\end{bmatrix}^T$ and $D_1 = \begin{bmatrix}0 & \sqrt{R}\end{bmatrix}^T$, then we get

$$\|G(s)\|_2^2 = \int_0^\infty (x(t)^T Q x(t) + u(t)^T R u(t)) \, dt, \tag{3.11}$$

which is the way the LQR problem is formulated.

3.2 Constrained $\mathscr{H}_2$ Optimization Algorithm

Given an LTI system (3.1a), (3.1b) and the control law (3.2), with $x(0) = x_0$, where $x \in \mathbb{R}^n$ is the state variable, $u \in \mathbb{R}^m$ is the control input, $w \in \mathbb{R}^l$ is the exogenous disturbance vector, $z \in \mathbb{R}^q$ is the controlled output. Matrices $A \in \mathbb{R}^{n\times n}$, $B_2 \in \mathbb{R}^{n\times m}$, $B_1 \in \mathbb{R}^{n\times l}$, $C_1 \in \mathbb{R}^{q\times n}$, $D_1 \in \mathbb{R}^{q\times m}$ are assumed to be known. The objective is to find the control gain $K \in \mathbb{R}^{m\times n}$ with certain constraints such that the following objective function is minimized, where

$$J = \int_0^\infty z(t)^T z(t) \, dt. \tag{3.12}$$

Notice that C_1 and D_1 are assumed to have appropriate dimensions such that $C_1^T D_1 = 0$ and $D_1^T D_1 > 0$.

The transfer function from w to z is given by

$$H(s) = (C_1 + D_1 K)(sI - A_c)^{-1} B_1, \tag{3.13}$$

where $A_c = A + B_2 K$. Assume $w(t) = \delta(t)$; by Parseval's Theorem, the optimization problem (3.12) is equivalent to minimizing the $\mathscr{H}_2$-norm of $H(s)$, yielding

$$\begin{aligned} J(K) &= \|H(s)\|_2^2 \\ &= \mathrm{Tr}\big((C_1 + D_1 K) W_c (C_1 + D_1 K)^T\big) \\ &= \mathrm{Tr}(B_1^T W_o B_1), \end{aligned} \tag{3.14}$$

where W_c and W_o are controllability and observability Gramians associated with the closed-loop system satisfying the following two equations

$$A_c W_c + W_c A_c^T + B_1 B_1^T = 0, \tag{3.15}$$

$$A_c^T W_o + W_o A_c + Q_c = 0, \tag{3.16}$$

where $Q_c = C_1^T C_1 + K^T D_1^T D_1 K$.

The equality constraints on the gain matrix K can be written as $K \in \Phi$, where $\Phi = \{K \in \mathbb{R}^{m \times n} : \mathcal{C}(K) = \mathcal{C}_0\}$, where the linear functions can be written as (2.14). Besides the equality constraints, the inequality constraint (2.15) is applied to keep the stability of the closed-loop system. Hence, the optimization problem is targeted to minimize $J(K)$ in (3.14) under constraints (2.14) and (2.15).

If no constraint is imposed on the feedback controller, K can be directly solved by the explicit formulae to design of the standard $\mathscr{H}_2$ optimal controller. However, due to the existence of equality constraint set (2.14), K cannot be directly computed. So an algorithm is required to obtain the optimal controller and flexure stiffness parameters iteratively while maintaining the stability condition (2.15).

Similarly, we consider the equality constraint (2.14) first; then the optimization problem is written in the short form as

$$\min_{K \in \Phi} J(K). \tag{3.17}$$

Define $\mathcal{C}_i(D) = 0$ as the hyperplane for the equality constraints (2.14), so (3.17) is converted to

$$\min_{D} \frac{1}{2} \left\| \frac{dJ(K)}{dK} - D \right\|^2, \quad s.t. \mathcal{C}_i(D) = 0, \quad \forall i = 1, \dots, N. \tag{3.18}$$

Problem (3.18) is convex, then it is further converted to the dual problem as

$$\max_{\Lambda_i} \min_{D} \left(\frac{1}{2} \left\| \frac{dJ(K)}{dK} - D \right\|^2 + \mathrm{Tr}\left(\sum_{i=1}^{N} \Lambda_i^T \mathcal{C}_i(D) \right) \right), \tag{3.19}$$

where $\Lambda_1, \dots, \Lambda_N$ are the dual variables associated with the equality constraints in (2.14), i.e. $\mathcal{C}_1(D) = 0$, ..., $\mathcal{C}_N(D) = 0$, respectively.

Remarkably, we can still use the Theorem 2.1 to ensure the monotonically decreasing of the cost.

The unconstrained gradient matrix $dJ(K)/dK$ determines a feasible direction of progress with a decreasing cost. The solutions in either finite horizon or infinite horizon are given by Theorem 3.1 and Theorem 3.2, respectively.

Theorem 3.1 *For the $\mathscr{H}_2$ control problem in the finite horizon ($T < \infty$),*

$$\frac{dJ(K)}{dK} = 2 \int_0^T (D_1^T D_1 K + B_2^T P(t)) X(t)\, dt, \tag{3.20}$$

with the following two stationarity conditions

$$A_c^T P(t) + P(t) A_c + Q_c = -\dot{P}(t), \tag{3.21}$$

$$A_c X(t) + X(t) A_c^T + B_1 w(t) x(t)^T + x(t) w(t)^T B_1^T = \dot{X}(t). \tag{3.22}$$

Proof of Theorem 3.1:

In the finite horizon, the cost function is given by

$$J(K) = \int_0^T z(t)^T z(t)\, dt = \mathrm{Tr}\left(\int_0^T Q_{\mathrm{c}} X(t)\, dt \right), \tag{3.23}$$

where $X(t) = x(t)x(t)^T$. From (3.23), we have

$$J(K) = \int_0^T f(X(t), K)\, dt, \tag{3.24}$$

where

$$f(X(t), K) = \mathrm{Tr}(Q_{\mathrm{c}} X(t)), \tag{3.25}$$

$$\dot{X}(t) = A_{\mathrm{c}} X(t) + X(t) A_{\mathrm{c}}^T + B_1 w(t) x(t)^T + x(t) w(t)^T B_1^T. \tag{3.26}$$

Thus,

$$\begin{aligned} H(X(t), P(t), K) = \mathrm{Tr}(Q_{\mathrm{c}} X(t)) + \mathrm{Tr}(P(t)^T (A_{\mathrm{c}} X(t) + X(t) A_{\mathrm{c}}^T \\ + B_1 w(t) x(t)^T + x(t) w(t)^T B_1^T)). \end{aligned} \tag{3.27}$$

We have

$$\frac{\partial}{\partial K} \mathrm{Tr}(Q_{\mathrm{c}} X(t)) = 2 D_1^T D_1 K X(t), \tag{3.28}$$

$$\begin{aligned} \frac{\partial}{\partial K} \mathrm{Tr}(P(t)^T (A_{\mathrm{c}} X(t) + X(t) A_{\mathrm{c}}^T + B_1 w(t) x(t)^T + x(t) w(t)^T B_1^T)) \\ = 2(P(t) B_2)^T X(t)^T. \end{aligned} \tag{3.29}$$

Then, the matrical gradient of the functional cost is given by

$$\begin{aligned} \frac{dJ(K)}{dK} &= \int_0^T \frac{\partial}{\partial K} H(X(t), P(t), K)\, dt \\ &= 2 \int_0^T (D_1^T D_1 K + B_2^T P(t)) X(t)\, dt, \end{aligned} \tag{3.30}$$

with boundary conditions (3.21) to (3.22). This concludes the proof of Theorem 3.1.

Theorem 3.2 *For the $\mathscr{H}_2$ control problem in the infinite horizon ($T = \infty$),*

$$\frac{dJ(K)}{dK} = 2(D_1^T D_1 K + B_2^T W_{\mathrm{o}}) W_{\mathrm{c}}. \tag{3.31}$$

Proof of Theorem 3.2: In this part, we extend the results in Theorem 3.1 to the infinite horizon ($T \to \infty$).

For (3.21), it is easy to extend to $T \to \infty$ because $\dot{P}(t) \to 0$ as $T \to \infty$, we have

$$A_c^T W_o + W_o A_c + Q_c = 0, \tag{3.32}$$

with $P(t) = W_o$ in the infinite horizon.

To extend (3.22) to the infinite horizon, first, we have

$$x(t) = \int_0^\infty e^{A_c(t-\tau)} B_1 w(\tau)\, d\tau + e^{A_c t} x_0. \tag{3.33}$$

To remove the initial state dependency, we have the following assumption.

Assumption 3.1 *In this $\mathscr{H}_2$ formulation, the mean of x_0 is zero.*

Under Assumption 3.1, $x(t) = \int_0^\infty e^{A_c(t-\tau)} B_1 w(\tau)\, d\tau$, and $X_0 = B_1 B_1^T$ with B_1 being a full rank matrix, which is tallied with the results in (Geromel, Bernussou, and Peres 1994). From the sifting property of $w(t) = \delta(t)$, we have

$$\int_0^\infty X(t)\, dt = \int_0^\infty e^{A_c t} B_1 B_1^T e^{A_c^T t}\, dt = W_c. \tag{3.34}$$

Take the integral of (3.21) on both sides from $t = 0$ to ∞, and substitute (3.34) in, and we have

$$A_c W_c + W_c A_c^T + \int_0^\infty B_1 w(\tau) x(t)^T + x(t) w(\tau)^T B_1^T\, d\tau = X(t) - X_0. \tag{3.35}$$

Therefore,

$$A_c W_c + W_c A_c^T + B_1 x(t)^T + x(t) B_1^T = X(t) - X_0. \tag{3.36}$$

As $t \to \infty$, we have $A_c W_c + W_c A_c^T + X_0 = 0$.

In this way, boundary conditions (3.21) to (3.22) are simplified to two Lyapunov equations associated with the observability and controllability Gramians of the closed-loop system as in (3.16) and (3.15). And the matrical gradient (3.20) is simplified to (3.31). This concludes the proof of Theorem 3.2.

The projection gradient matrix onto the equality constrained hyperplane can be obtained using Theorem 2.4. We summarize the above discussion into the following algorithm.

Theorem 3.3 is used to show that both the equality constraints (2.14) and the inequality constraint (2.15) will be satisfied for optimized K^i for all $i \geq 1$ if certain conditions are met.

Theorem 3.3 *Regarding K^i with constraints* (2.14) *and* (2.15), *the following statements hold.*

1. *If K^0 is designed to satisfy the equality constraints in* (2.14), *so will K^i be for all $i \geq 1$.*

Algorithm 3.1 Constrained $\mathscr{H}_2$ Optimization Algorithm

- Step 1: Set $i = 0$ and initialize a stable gain $K^0 \in \Phi$.
- Step 2: Set $i = i + 1$ and determine the unconstrained gradient matrix $(dJ(K)/dK)^i$ by Theorem 3.2.
- Step 3: Determine the projection gradient matrix D^i by Theorem 2.4.
- Step 4: Optimize the step size α^i to minimize the functional cost after the iteration, which is a line search problem written as $\min_{\alpha^i} J(K^{i-1} - \alpha^i D^i)$.
- Step 5: Gain matrix is updated as $K^i = K^{i-1} - \alpha^i D^i$.
- Step 6: Go back to Step 2 to continue the iterations until reaching the stopping criterion.

---

2. *K^i will satisfy the inequality constraint in* (2.15) *for all $i \geq 1$ if*
    - *The pair (A, F) is detectable with $C_1^T C_1 = F^T F$;*
    - *(A, H_0) is stabilizable with $X_0 = H_0 H_0^T$, where $H_0 \in \mathbb{R}^{n \times s}$ is a full rank matrix;*
    - $(I_n - H_0(H_0^T H_0)^{-1} H_0^T)B_2 = 0$.

Proof of Theorem 3.3: The proof of Statement 1 of Theorem 3.3 is the same as in Theorem 2.5.

To prove Statement 2, first, we make a transformation of the cost function in the infinite horizon, which is given by

$$J(K) = \mathrm{Tr}(W_o X_0) = \mathrm{Tr}\left(\int_0^\infty Q_c e^{A_c t} B_1 B_1^T e^{A_c^T t}\, dt\right). \tag{3.37}$$

From (3.37), we want to prove $\lim_{t\to\infty} e^{A_c t} B_1 = 0$ is equivalent to $0 < J(K) < \infty$: if $B_1 \in \mathrm{span}(H_0)$, $\lim_{t\to\infty} e^{A_c t} B_1 = 0$ is equivalent to $\lim_{t\to\infty} e^{A_c t} h_{oi} = 0$, where h_{oi} is the ith column of H_0, $i = 1, \ldots, s$. It is also equivalent to $0 < J(K) = \mathrm{Tr}(W_o X_0) = \sum_{n=1}^{s} h_{oi}^T W_o h_{oi} < \infty$ due to the detectability of (A, F). The remaining proof is the same as the proof in Theorem 2.5.

3.3 Case Study

For simplification, "(t)" is omitted in the expression of time-history signals in this section.

3.3.1 Statement of Problem

DHG owns the advantage of high-power density on the principle axis due to its dual-drive motors, and it yields high-speed motion with no significant lateral offset when two carriages are appropriately coordinated and synchronized in motion. However, various factors on the whole system may cause the desynchronization of two carriages, such as non-uniform load distribution of the DHG, mismatch of motors and drive characteristics, as well as time-varying thermal-mechanical properties (Tan, Lim, Huang, Dou, and Giam 2004). Efficient synchronization of these two carriages via either controller or mechanical design is crucially important to minimize the inter-axis error and maintain the orthogonality between X and Y axes. Some trials on further improvement of the synchronization are done by introducing the limited information exchange between two parallel axes, such as cross-coupled control schemes that utilize adaptive parameter estimation (Teo, Tan, Lim, Huang, and Tay 2007), and the use of DOB (Li, Chen, Teo, and Tan 2017; Chen, Li, Teo, and Tan 2017) for synchronous motion. However, the quantitative modeling of the inter-axis coupling force is still absent in such pure controller design approaches.

Classically, the biaxial synchronization can be enhanced by imposing the rigid joints at both ends of the cross-arm (Giam, Tan, and Huang 2007; Yao 2015), but the joints may be damaged as discussed in Section 2.3.1. Hence, the flexure-linked design is suggested, while the resonant mode suppression needs to be considered. For this flexure-linked DHG, the assurance of the biaxial synchronization and orthogonality in the flexure-linked DHG requires simultaneous determination of the controller parameters and the stiffness of the flexure joints.

In this section, symbols for all parameters used in this work are demonstrated in Table 3.1. In our experiments, we only initiate our design based on the motion of the Y-axis, while holding the X-axis actuator at the center. Here, we consider the Coulomb friction as a step disturbance and we group the viscous friction into the damping coefficients of the system. In Figure 3.1(a), the schematic diagram of the DHG is demonstrated, which can be simplified as a two-mass model linked by a pair of flexible joints with a light rod; the simplified schematic diagram is illustrated in Figure 3.1(b). Thus, the DHG system is now modeled as a coupled linear system

$$M_1\ddot{y}_1 = K_f u_1 - \Gamma_1\dot{y}_1 + v - f_1 w, \tag{3.38a}$$
$$M_2\ddot{y}_2 = K_f u_2 - \Gamma_2\dot{y}_2 - v - f_2 w, \tag{3.38b}$$

where $M_1 = (m_e + m_c)/2 + m_1$ and $M_2 = (m_e + m_c)/2 + m_2$, and w is a unit step function. Here, we assume that the damping coefficients of two axes are the same, where $\Gamma_1 = \Gamma_2 = \Gamma$.

In this work, two carriages are targeted to track an identical autonomous S-curve trajectory, with minimizing the possible induced chattering from flexure

Table 3.1: Nomenclature used in Chapter 3 © [2017] Elsevier. Reprinted, with permission, from J. Ma, S.-L. Chen, N. Kamaldin, C. S. Teo, A. Tay, A. Al. Mamun, and K. K. Tan, A Novel Constrained H_2 Optimization Algorithm for Mechatronics Design in Flexure-Linked Biaxial Gantry, ISA Transactions, vol. 71, pp. 467–479, 2017.

Name	Unit	Description
K_f	N/A	Force constant
m_1	kg	Mass of carriage 1
m_2	kg	Mass of carriage 2
m_e	kg	Mass of end effector
m_c	kg	Mass of cross-arm
Γ_1	Ns/m	Damping coefficient in carriage 1
Γ_2	Ns/m	Damping coefficient in carriage 2
f_1	N	Coulomb friction in carriage 1
f_2	N	Coulomb friction in carriage 2
y_1	m	Position of carriage 1
y_2	m	Position of carriage 2
y_d	m	Reference position for two carriages
e_i	m	Inter-axis error
u_1	A	Control current of carriage 1
u_2	A	Control current of carriage 2
v	N	Coupling force from flexure
k_v	N/m	Stiffness of flexure

force and control signals. Equivalently, these requirements are quantified as

- The master axis y_1 tracks the S-curve trajectory $y_\mathrm{d} = p_1$, where $\dot{p} = A_\mathrm{p}p$, $p^T = \begin{bmatrix} p_1 & p_2 & p_3 \end{bmatrix}$, $A_\mathrm{p} = \begin{bmatrix} 0 & 1 & 0 \\ 0 & 0 & 1 \\ a_\mathrm{p1} & a_\mathrm{p2} & a_\mathrm{p3} \end{bmatrix}$.
- The inter-axis error between two axes $e_\mathrm{i} = y_1 - y_2$ is to be kept small.
- The chattering of control signals $\dot{u}_1$, $\dot{u}_2$ and the jerk of the coupling force from the flexure $\dot{v}$ should be kept small to avoid activating resonance induced by the flexure joints.

3.3.2 Formulation of Constrained $\mathscr{H}_2$ Optimization Problem

The system model in (3.38) is standardized to the following state-space form

$$\dot{x} = Ax + B_2u + B_1w, \tag{3.39}$$

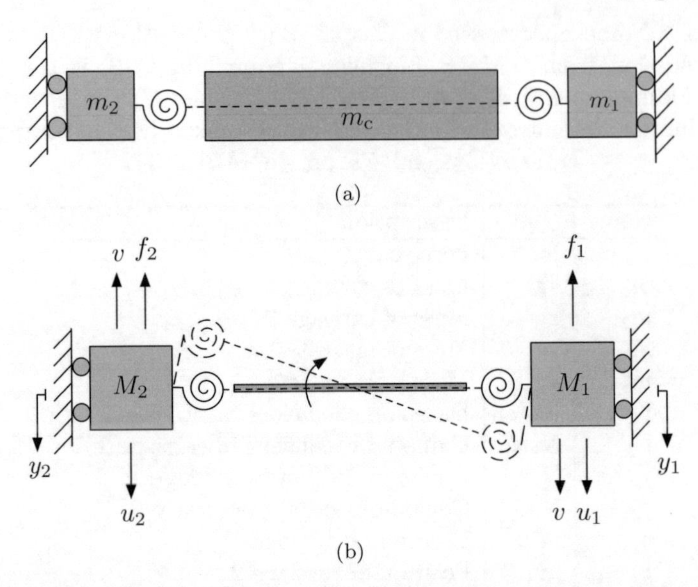

Figure 3.1: Schematic of DHG without X-axis carriage. (a) Actual model. (b) Simplified model.

where $x^T = \begin{bmatrix} y_1 & \dot{y}_1 & y_2 & \dot{y}_2 \end{bmatrix}$, $u^T = \begin{bmatrix} u_1 & u_2 & v \end{bmatrix}$. Matrices A, B_2 and B_1 are given by

$$A = \begin{bmatrix} 0 & 1 & 0 & 0 \\ 0 & -\frac{\Gamma_1}{M_1} & 0 & 0 \\ 0 & 0 & 0 & 1 \\ 0 & 0 & 0 & -\frac{\Gamma_2}{M_2} \end{bmatrix},$$

$$B_2 = \begin{bmatrix} 0 & 0 & 0 \\ \frac{K_f}{M_1} & 0 & \frac{1}{M_1} \\ 0 & 0 & 0 \\ 0 & \frac{K_f}{M_2} & -\frac{1}{M_2} \end{bmatrix}, \quad B_1 = \begin{bmatrix} 0 \\ -\frac{f_1}{M_1} \\ 0 \\ -\frac{f_2}{M_2} \end{bmatrix}. \tag{3.40}$$

Since we will follow the industrial practice to use the PID controller for the DHG, (3.39) is differentiated and augmented with y_1, y_2 as new state vectors, yielding

$$\begin{aligned} \hat{x}^T &= \begin{bmatrix} y_1 & \dot{y}_1 & \ddot{y}_1 & y_2 & \dot{y}_2 & \ddot{y}_2 \end{bmatrix} \\ &= \begin{bmatrix} \hat{x}_1 & \hat{x}_2 & \hat{x}_3 & \hat{x}_4 & \hat{x}_5 & \hat{x}_6 \end{bmatrix}. \end{aligned} \tag{3.41}$$

Notably, $\dot{w}$ is an impulse function. The extended state-space representation of the system is given by

$$\dot{\hat{x}} = \hat{A}\hat{x} + \hat{B}_2\dot{u} + \hat{B}_1\dot{w}, \tag{3.42}$$

where matrices $\hat{A}$, $\hat{B}_2$ and $\hat{B}_1$ are given by

$$\hat{A} = \begin{bmatrix} 0 & 1 & 0 & 0 & 0 & 0 \\ 0 & 0 & 1 & 0 & 0 & 0 \\ 0 & 0 & -\frac{\Gamma_1}{M_1} & 0 & 0 & 0 \\ 0 & 0 & 0 & 0 & 1 & 0 \\ 0 & 0 & 0 & 0 & 0 & 1 \\ 0 & 0 & 0 & 0 & 0 & -\frac{\Gamma_2}{M_2} \end{bmatrix},$$

$$\hat{B}_2 = \begin{bmatrix} 0 & 0 & 0 \\ 0 & 0 & 0 \\ \frac{K_f}{M_1} & 0 & \frac{1}{M_1} \\ 0 & 0 & 0 \\ 0 & 0 & 0 \\ 0 & \frac{K_f}{M_2} & -\frac{1}{M_2} \end{bmatrix}, \quad \hat{B}_1 = \begin{bmatrix} 0 \\ 0 \\ -\frac{f_1}{M_1} \\ 0 \\ 0 \\ -\frac{f_2}{M_2} \end{bmatrix}. \tag{3.43}$$

We augment the reference trajectory with the system (3.42), yielding

$$\dot{\bar{x}} = \bar{A}\bar{x} + \bar{B}_2\dot{\bar{u}} + \bar{B}_1\dot{\bar{w}}, \tag{3.44}$$

where $\bar{x}^T = \begin{bmatrix} p^T & \hat{x}^T \end{bmatrix}$, $\bar{A} = \text{blocdiag}\{A_\text{p}, \hat{A}\}$, $\bar{B}_2^T = \begin{bmatrix} 0_{3\times 3} & \hat{B}_2^T \end{bmatrix}$, $\bar{B}_1^T = \begin{bmatrix} 0_{1\times 3} & \hat{B}_1^T \end{bmatrix}$, $\bar{u} = u$ and $\bar{w} = w$.

Remark 3.1 *If* $y_\text{d} = p_1 + y_\Delta$, *where* y_Δ *is the shift on* p_1 *for a zero starting position, then we can set* $\hat{x}^T = \begin{bmatrix} y_1 - y_\Delta & \dot{y}_1 & \ddot{y}_1 & y_2 - y_\Delta & \dot{y}_2 & \ddot{y}_2 \end{bmatrix}$ *for* (3.42), *such that* $e_1 = y_\text{d} - y_1 = p_1 - \hat{x}_1$, $e_2 = y_\text{d} - y_2 = p_1 - \hat{x}_4$.

Based on the optimization targets, the objective function is defined as

$$J = \int_0^\infty z^T z \, dt, \tag{3.45}$$

where z is the controlled output expressed as

$$z = C_\text{z} \begin{bmatrix} y_\text{d} - y_1 & y_1 - y_2 & \dot{u}_1 & \dot{u}_2 & \dot{v} \end{bmatrix}^T, \tag{3.46}$$

where $C_\text{z} = \text{diag}\{q_{\text{c}1}, q_{\text{c}2}, q_{\text{d}1}, q_{\text{d}2}, q_{\text{d}3}\}$.

Equivalently, z can be expressed as $z = \bar{C}\bar{x} + \bar{D}\dot{\bar{u}}$, where $\bar{C} = Q_\text{C}\hat{C}$, $\bar{D} = Q_\text{D}\hat{D}$, $\hat{C} = \begin{bmatrix} 1 & 0 & 0 & -1 & 0 & 0 & 0 & 0 & 0 \\ 0 & 0 & 0 & 1 & 0 & 0 & -1 & 0 & 0 \\ & & & & 0_{3\times 9} & & & & \end{bmatrix}$, $\hat{D} = \begin{bmatrix} 0_{2\times 3} \\ I_3 \end{bmatrix}$, and $Q_\text{C} = \text{diag}\{q_{\text{c}1}, q_{\text{c}2}, 0, 0, 0\}$, $Q_\text{D} = \text{diag}\{0, 0, q_{\text{d}1}, q_{\text{d}2}, q_{\text{d}3}\}$. $q_{\text{c}1}$, $q_{\text{c}2}$, $q_{\text{d}1}$, $q_{\text{d}2}$, $q_{\text{d}3}$ are user-defined values representing penalties on different factors in the objective function, including tracking performance of carriage 1, synchronization between two axes as well as chattering of control signals and flexure force.

The transfer function from $\dot{\bar{w}}$ to z is given by

$$H(s) = (\bar{C} + \bar{D}K)(sI - A_c)^{-1}\bar{B}_1, \tag{3.47}$$

where $A_c = \bar{A} + \bar{B}K$. The $\mathscr{H}_2$-norm is interpreted against deterministic disturbance input. By Parseval's Theorem, the optimization problem (3.45) is equivalent to minimizing the $\mathscr{H}_2$-norm of $H(s)$, yielding

$$\begin{aligned} J(K) &= \|H(s)\|_2^2 \\ &= \mathrm{Tr}\big((\bar{C} + \bar{D}K)W_c(\bar{C} + \bar{D}K)^T\big) \\ &= \mathrm{Tr}(\bar{B}_1^T W_o \bar{B}_1), \end{aligned} \tag{3.48}$$

where W_c and W_o are controllability and observability Gramians associated with the closed-loop system satisfying the following two equations

$$A_c W_c + W_c A_c^T + \bar{B}_1 \bar{B}_1^T = 0, \tag{3.49}$$

$$A_c^T W_o + W_o A_c + Q_c = 0. \tag{3.50}$$

where $A_c = \bar{A} + \bar{B}K$, $Q_c = \bar{C}^T\bar{C} + K^T\bar{D}^T\bar{D}K$.

The generalized feedback controller is given by

$$\dot{\bar{u}} = K\bar{x}. \tag{3.51}$$

Since there are only two physical control inputs u_1 and u_2 in the actual system, we split the controller gain K as

$$K = \begin{bmatrix} K_F & K_B \\ 0_{1\times 3} & K_V \end{bmatrix} \in \mathbb{R}^{3\times 9}, \tag{3.52}$$

with

$$K_F = \begin{bmatrix} k_{z1} & k_{z2} & k_{z3} \\ k_{z4} & k_{z5} & k_{z6} \end{bmatrix}, \tag{3.53}$$

$$K_B = \mathrm{blocdiag}\{\, [-k_{i1} \ -k_{p1} \ -k_{d1}]\,, [-k_{i2} \ -k_{p2} \ -k_{d2}] \,\}, \tag{3.54}$$

$$K_V = [0 \quad k_{s1} \quad 0 \quad 0 \quad k_{s2} \quad 0]\,. \tag{3.55}$$

In reality, the coupling force from the flexure is

$$v = k_v(y_1 - y_2). \tag{3.56}$$

Thus, an equality constraint is naturally given as

$$k_{s1} = -k_{s2} = -k_v. \tag{3.57}$$

K_B is structured as a block diagonal matrix due to the decentralized nature of u_1 and u_2. In addition, the control structures for both sides of parallel axes are constrained to be the standard PID forms, where k_{p1}, k_{i1} and k_{d1} represent the proportional gain, the integral gain and the derivative gain, respectively, and the same for k_{p2}, k_{i2} and k_{d2}. Thus, additional equality constraints are given as

$$k_{z1} = k_{i1},\ k_{z2} = k_{p1},\ k_{z3} = k_{d1},\ k_{z4} = k_{i2},\ k_{z5} = k_{p2},\ k_{z6} = k_{d2}. \tag{3.58}$$

Equivalently, (3.57) and (3.58) are written as $K \in \Phi$, where $\Phi = \{K \in \mathbb{R}^{3\times 9} : \mathcal{C}(K) = \mathcal{C}_0\}$, and $\mathcal{C}(K)$ is a function of K with $\mathcal{C}(0) = 0$. Then , it is easy to show that the constraints (3.57) and (3.58) are equivalent to

$$\mathcal{C}_1(K) = \tilde{A}_1 K \tilde{B}_1 + \tilde{A}_2 K \tilde{B}_2 + \tilde{A}_3 K \tilde{B}_3 = 0, \tag{3.59a}$$

$$\mathcal{C}_2(K) = \tilde{A}_4 K \tilde{B}_4 + \tilde{A}_5 K \tilde{B}_5 = 0, \tag{3.59b}$$

$$\mathcal{C}_3(K) = \tilde{A}_6 K \tilde{B}_6 + \tilde{A}_7 K \tilde{B}_7 = 0, \tag{3.59c}$$

$$\mathcal{C}_4(K) = \tilde{A}_8 K \tilde{B}_8 + \tilde{A}_9 K \tilde{B}_9 = 0, \tag{3.59d}$$

$$\mathcal{C}_5(K) = \tilde{A}_{10} K \tilde{B}_{10} + \tilde{A}_{11} K \tilde{B}_{11} = 0, \tag{3.59e}$$

$$\mathcal{C}_6(K) = \tilde{A}_{12} K \tilde{B}_{12} + \tilde{A}_{13} K \tilde{B}_{13} = 0, \tag{3.59f}$$

$$\mathcal{C}_7(K) = \tilde{A}_{14} K \tilde{B}_{14} + \tilde{A}_{15} K \tilde{B}_{15} = 0, \tag{3.59g}$$

$$\mathcal{C}_8(K) = \tilde{A}_{16} K \tilde{B}_{16} + \tilde{A}_{17} K \tilde{B}_{17} = 0, \tag{3.59h}$$

where matrices $\tilde{A}_j$ and $\tilde{B}_j$ $(j = 1, \ldots, 17)$ are defined as

$$\begin{aligned}
\tilde{A}_1 &= \mathrm{diag}\{1,\ 0,\ 0\},\\
\tilde{A}_2 &= \mathrm{diag}\{0,\ 1,\ 0\},\\
\tilde{A}_3 &= \mathrm{diag}\{0,\ 0,\ 1\},\\
\tilde{A}_4 = \tilde{A}_5 = \tilde{A}_6 = \tilde{A}_7 = \tilde{A}_8 = \tilde{A}_9 &= \begin{bmatrix} 1 & 0 & 0 \end{bmatrix},\\
\tilde{A}_{10} = \tilde{A}_{11} = \tilde{A}_{12} = \tilde{A}_{13} = \tilde{A}_{14} = \tilde{A}_{15} &= \begin{bmatrix} 0 & 1 & 0 \end{bmatrix},\\
\tilde{A}_{16} = \tilde{A}_{17} &= \begin{bmatrix} 0 & 0 & 1 \end{bmatrix},\\
\tilde{B}_1 &= \mathrm{diag}\{0,\ 0,\ 0,\ 0,\ 0,\ 0,\ 1,\ 1,\ 1\},\\
\tilde{B}_2 &= \mathrm{diag}\{0,\ 0,\ 0,\ 1,\ 1,\ 1,\ 0,\ 0,\ 0\},\\
\tilde{B}_3 &= \mathrm{diag}\{1,\ 1,\ 1,\ 1,\ 0,\ 1,\ 1,\ 0,\ 1\},\\
\tilde{B}_4 = \tilde{B}_{10} &= \begin{bmatrix} 1 & 0 & 0 & 0 & 0 & 0 & 0 & 0 & 0 \end{bmatrix}^T,\\
\tilde{B}_5 &= \begin{bmatrix} 0 & 0 & 0 & 1 & 0 & 0 & 0 & 0 & 0 \end{bmatrix}^T,\\
\tilde{B}_6 = \tilde{B}_{12} &= \begin{bmatrix} 0 & 1 & 0 & 0 & 0 & 0 & 0 & 0 & 0 \end{bmatrix}^T,\\
\tilde{B}_7 = \tilde{B}_{16} &= \begin{bmatrix} 0 & 0 & 0 & 0 & 1 & 0 & 0 & 0 & 0 \end{bmatrix}^T,\\
\tilde{B}_8 = \tilde{B}_{14} &= \begin{bmatrix} 0 & 0 & 1 & 0 & 0 & 0 & 0 & 0 & 0 \end{bmatrix}^T,\\
\tilde{B}_9 &= \begin{bmatrix} 0 & 0 & 0 & 0 & 0 & 1 & 0 & 0 & 0 \end{bmatrix}^T,\\
\tilde{B}_{11} &= \begin{bmatrix} 0 & 0 & 0 & 0 & 0 & 0 & 1 & 0 & 0 \end{bmatrix}^T,\\
\tilde{B}_{13} = \tilde{B}_{17} &= \begin{bmatrix} 0 & 0 & 0 & 0 & 0 & 0 & 0 & 1 & 0 \end{bmatrix}^T,\\
\tilde{B}_{15} &= \begin{bmatrix} 0 & 0 & 0 & 0 & 0 & 0 & 0 & 0 & 1 \end{bmatrix}^T.
\end{aligned} \tag{3.60}$$

Besides, to ensure the stability of the closed-loop system, the following inequality constraint is defined as

$$\mathrm{Re}(\mathrm{eig}(A_c)) < 0. \tag{3.61}$$

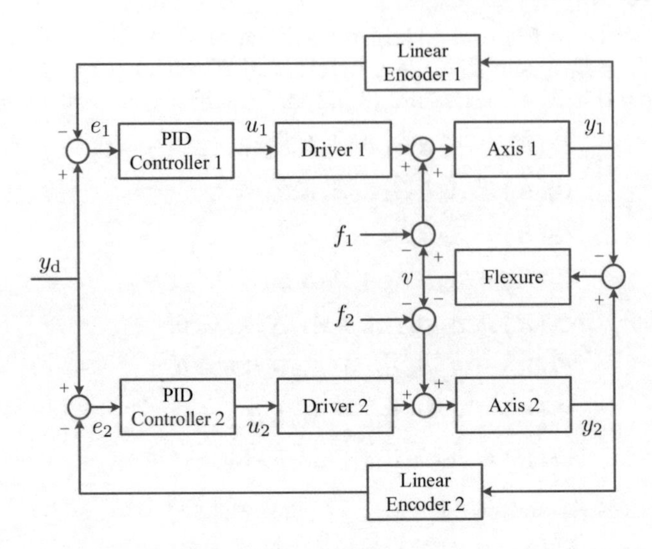

Figure 3.2: Overview of decentralized control architecture.

The objective is to minimize the cost function (3.48) under the constraints (3.59) and (3.61).

By constraint set (3.58), the standard PID controllers are restored as

$$u_1 = k_{p1}e_1 + k_{i1}\int_0^t e_1\, d\tau + k_{d1}\dot{e}_1, \tag{3.62a}$$

$$u_2 = k_{p2}e_2 + k_{i2}\int_0^t e_2\, d\tau + k_{d2}\dot{e}_2. \tag{3.62b}$$

Eventually, the block diagram of the overall closed-loop system is illustrated in Figure 3.2.

The optimization problem is converted to

$$\min_D \frac{1}{2}\left\|\frac{dJ(K)}{dK} - D\right\|^2, \quad s.t.\, \mathcal{C}_i(D) = 0, \quad \forall i = 1,\dots,8. \tag{3.63}$$

Then, (3.63) is converted to its equivalent dual problem as

$$\max_{\Lambda_i}\min_D \left(\frac{1}{2}\left\|\frac{dJ(K)}{dK} - D\right\|^2 + \mathrm{Tr}\left(\sum_{i=1}^{8}\Lambda_i^T\mathcal{C}_i(D)\right)\right), \tag{3.64}$$

where $\Lambda_1,\dots,\Lambda_8$ are the dual variables associated with the equality constraints in (3.59), i.e. $\mathcal{C}_1(D) = 0, \dots, \mathcal{C}_8(D) = 0$, respectively.

After obtaining $dJ(K)/dK$, we will project it onto $\mathcal{C}_i(D) = 0$ to get D, which can be determined by Corollary 3.1.

Corollary 3.1 *The projection gradient matrix is given by*

$$D = \frac{dJ(K)}{dK} - \sum_{j=1}^{3} \tilde{A}_j \frac{dJ(K)}{dK} \tilde{B}_j - \sum_{j=4}^{5} \tilde{A}_j^T \Lambda_2 \tilde{B}_j^T - \sum_{j=6}^{7} \tilde{A}_j^T \Lambda_3 \tilde{B}_j^T$$
$$- \sum_{j=8}^{9} \tilde{A}_j^T \Lambda_4 \tilde{B}_j^T - \sum_{j=10}^{11} \tilde{A}_j^T \Lambda_5 \tilde{B}_j^T - \sum_{j=12}^{13} \tilde{A}_j^T \Lambda_6 \tilde{B}_j^T$$
$$- \sum_{j=14}^{15} \tilde{A}_j^T \Lambda_7 \tilde{B}_j^T - \sum_{j=16}^{17} \tilde{A}_j^T \Lambda_8 \tilde{B}_j^T, \tag{3.65}$$

and the dual variable is given by

$$\Lambda_i = \frac{1}{2} \left(\tilde{A}_{2i} \frac{dJ(K)}{dK} \tilde{B}_{2i} + \tilde{A}_{2i+1} \frac{dJ(K)}{dK} \tilde{B}_{2i+1} \right), \tag{3.66}$$

where $i = 2, \ldots, 8$.

Proof of Corollary 3.1: The necessary and sufficient optimal solution to the dual problem (3.64) is given by

$$D - \frac{dJ(K)}{dK} + \frac{\partial}{\partial D} \left(\mathrm{Tr} \left(\sum_{i=1}^{8} \Lambda_j^T \mathcal{C}_i(D) \right) \right) = 0, \tag{3.67}$$

which leads to

$$D - \frac{dJ(K)}{dK} + \sum_{j=1}^{3} \tilde{A}_j^T \Lambda_1 \tilde{B}_j^T + \sum_{j=4}^{5} \tilde{A}_j^T \Lambda_2 \tilde{B}_j^T + \sum_{j=6}^{7} \tilde{A}_j^T \Lambda_3 \tilde{B}_j^T$$
$$+ \sum_{j=8}^{9} \tilde{A}_j^T \Lambda_4 \tilde{B}_j^T + \sum_{j=10}^{11} \tilde{A}_j^T \Lambda_5 \tilde{B}_j^T + \sum_{j=12}^{13} \tilde{A}_j^T \Lambda_6 \tilde{B}_j^T$$
$$+ \sum_{j=14}^{15} \tilde{A}_j^T \Lambda_7 \tilde{B}_j^T + \sum_{j=16}^{17} \tilde{A}_j^T \Lambda_8 \tilde{B}_j^T = 0. \tag{3.68}$$

By $\tilde{A}_1 \times (3.68) \times \tilde{B}_1 + \tilde{A}_2 \times (3.68) \times \tilde{B}_2 + \tilde{A}_3 \times (3.68) \times \tilde{B}_3$, we have

$$\sum_{j=1}^{3} \tilde{A}_j^T \Lambda_1 \tilde{B}_j^T = \sum_{j=1}^{3} \tilde{A}_j \frac{dJ(K)}{dK} \tilde{B}_j. \tag{3.69}$$

(3.66) can be shown in a similar way. Eventually, substitute (3.66) and (3.69) into (3.68), it gives (3.65). This proves Corollary 3.1.

3.3.3 Optimization and Experimental Validation

The proposed constrained $\mathscr{H}_2$ optimization algorithm is used for the integrated design of the controller and the flexure for the DHG. Similar to Chapter 2, we aim to find the most suitable flexure among those three pieces and determine the optimal controller parameters to minimize the objective function. Hence, we need to slightly modify the optimization strategy by fixing the value k_v to be the values of those of the three pieces and optimize the controllers' parameters subsequently, so we set Λ_8 to 0 and change $\tilde{B}_3$ to $\tilde{B}_{3+}$, where $\tilde{B}_{3+} = \text{diag}\{1, 1, 1, 1, 1, 1, 1, 1, 1\}$. For every piece of flexure joint with different thickness, we set a number of 1000 iterations as the stopping criterion of the algorithm and compare the costs to see which flexure gives the lowest one. Eventually, we carry on the motion experiments using the optimized parameters.

Coulomb frictions in two carriages are identified as $f_1 = 0.1193$ N, $f_2 = 0.1544$ N, respectively, using a relay tuning method outlined in (Chen, Tan, Huang, and Teo 2010), and the viscous frictions are not independently identified because they can be grouped into the damping terms when we do the system identification for the DHG. The eigenvalues of matrix A_p are designed to obtain a required S-curve profile; in this work, we set the all of the eigenvalues to -4, then a_{p1}, a_{p2}, a_{p3} are obtained as -64, -48 and -12, respectively. The DHG is aimed to move -0.1 m along the axes, so we initiate the position of the S-curve profile to be started at -0.1 m and then shift up the generated profile by 0.1 m to allow the profile to start at zero. The motion profile is shown in Figure 3.3. As in Remark 3.1, the optimization is based on the certain shape of S-curve profile determined by A_p, hence, there is no influence on the optimization results if we shift the S-curve profile or do the mirror flip of it up and down.

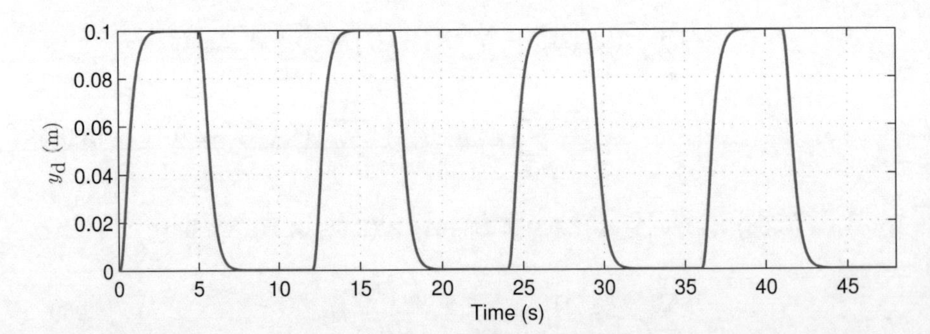

Figure 3.3: Reference profile for both axes. © [2017] Elsevier. Reprinted, with permission, from J. Ma, S.-L. Chen, N. Kamaldin, C. S. Teo, A. Tay, A. Al. Mamun, and K. K. Tan, A Novel Constrained H_2 Optimization Algorithm for Mechatronics Design in Flexure-Linked Biaxial Gantry, ISA Transactions, vol. 71, pp. 467–479, 2017.

Different penalties can be set based on different requirements on the performance. The weightings are quantified based on the selection of $q_{c1}, q_{c2}, q_{d1}, q_{d2}, q_{d3}$.

We set the weightings in the cost function to be $q_{c1} = 10^4, q_{c2} = 10^4, q_{d1} = 50, q_{d2} = 50, q_{d3} = 50$. We define the initial gain matrix in the proposed optimization as

$$K_{\mathrm{B}}^{0} = \mathrm{blocdiag}\{ [-50 - 600 - 10], [-50 - 600 - 10] \}. \tag{3.70}$$

With the 2-mm, the 3-mm, and the 4-mm flexure joints, the initial gain matrices stabilize the closed-loop system with functional costs of 30.77, 82.21 and 164.55, respectively. We run the optimization algorithm for 1000 iterations, and their costs are decreasing significantly, which are 4.67, 6.61 and 8.54, respectively. We choose and install the 2-mm flexure for the following-up motion control experiments because it results in the lowest value of the functional cost.

Figure 3.4 shows the functional cost J and the norm of the projection gradient matrix $\|D\|$ using three flexures within the first 5 iterations. It is observed that the functional cost is decreasing with iterations. Notice that

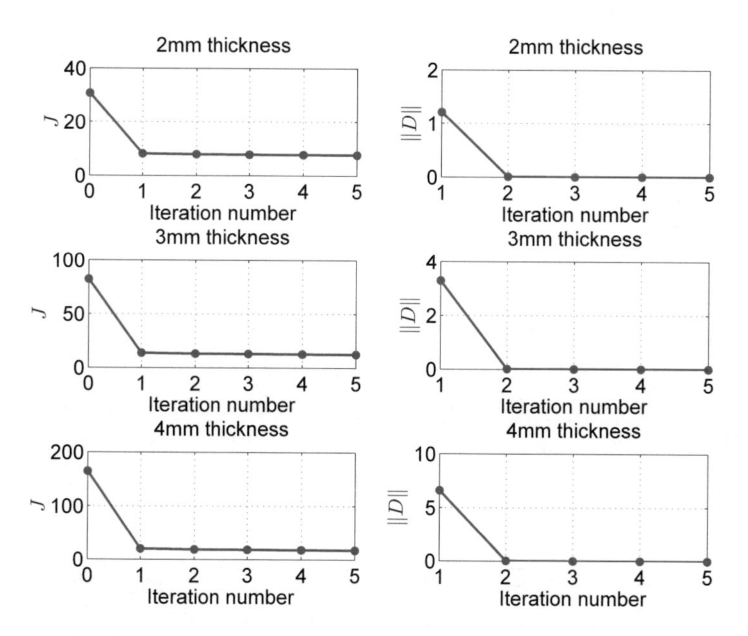

Figure 3.4: Functional cost and norm of projection gradient matrix during iterations. © [2017] Elsevier. Reprinted, with permission, from J. Ma, S.-L. Chen, N. Kamaldin, C. S. Teo, A. Tay, A. Al. Mamun, and K. K. Tan, A Novel Constrained H_2 Optimization Algorithm for Mechatronics Design in Flexure-Linked Biaxial Gantry, ISA Transactions, vol. 71, pp. 467–479, 2017.

the cost is decreasing most significantly in the first iteration. For the following iterations, it is still decreasing, but with much smaller rates.

The comparative experiments are conducted using the initial gain, the optimized gain with 100 iterations as well as the optimized gain with 1000 iterations, where the last two feedback gains obtained from optimization are given by

$$\begin{aligned} K_{\mathrm{B}}^{100} = \mathrm{blocdiag}\{ & \left[-84.1 - 1168.4 - 67.9\right], \\ & \left[-77.6 - 1361.4 - 67.4\right] \}, \end{aligned} \tag{3.71}$$

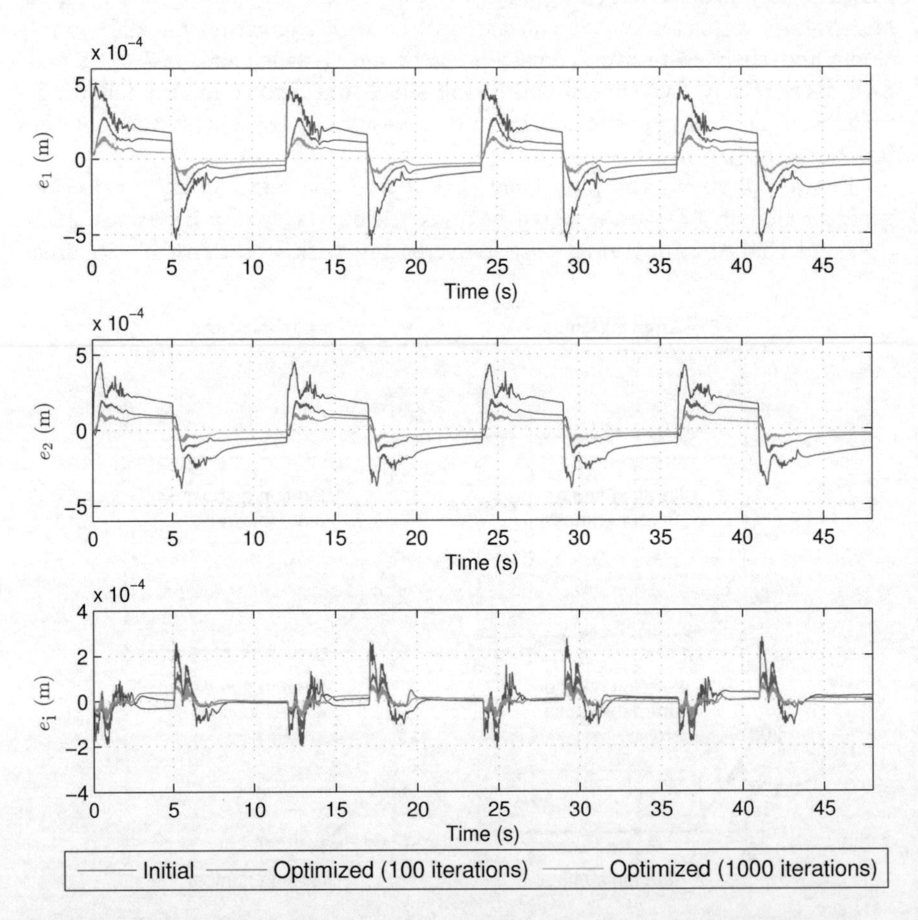

Figure 3.5: Comparison of tracking errors of two axes and inter-axis error with different iteration numbers. © [2017] Elsevier. Reprinted, with permission, from J. Ma, S.-L. Chen, N. Kamaldin, C. S. Teo, A. Tay, A. Al. Mamun, and K. K. Tan, A Novel Constrained H_2 Optimization Algorithm for Mechatronics Design in Flexure-Linked Biaxial Gantry, ISA Transactions, vol. 71, pp. 467–479, 2017.

$$K_{\mathrm{B}}^{1000} = \mathrm{blocdiag}\{ \left[-122.7 \ -2237.1 \ -53.7\right], \\ \left[-108.8 \ -2600.0 \ -54.4\right] \}. \tag{3.72}$$

Figure 3.5 shows the tracking errors of both carriages as well as the inter-axis error with different iteration numbers, i.e. $i = 0, 100, 1000$. For validation purposes, we use the laser interferometer to measure the rotation angle of the end effector along the cross-arm. This reading should be proportional to the inter-axis error if the link bar is not vibrating during motion. The comparison of inter-axis error calculated based on laser interferometer reading and the encoder is illustrated in Figure 3.6, showing the same trends for all the three

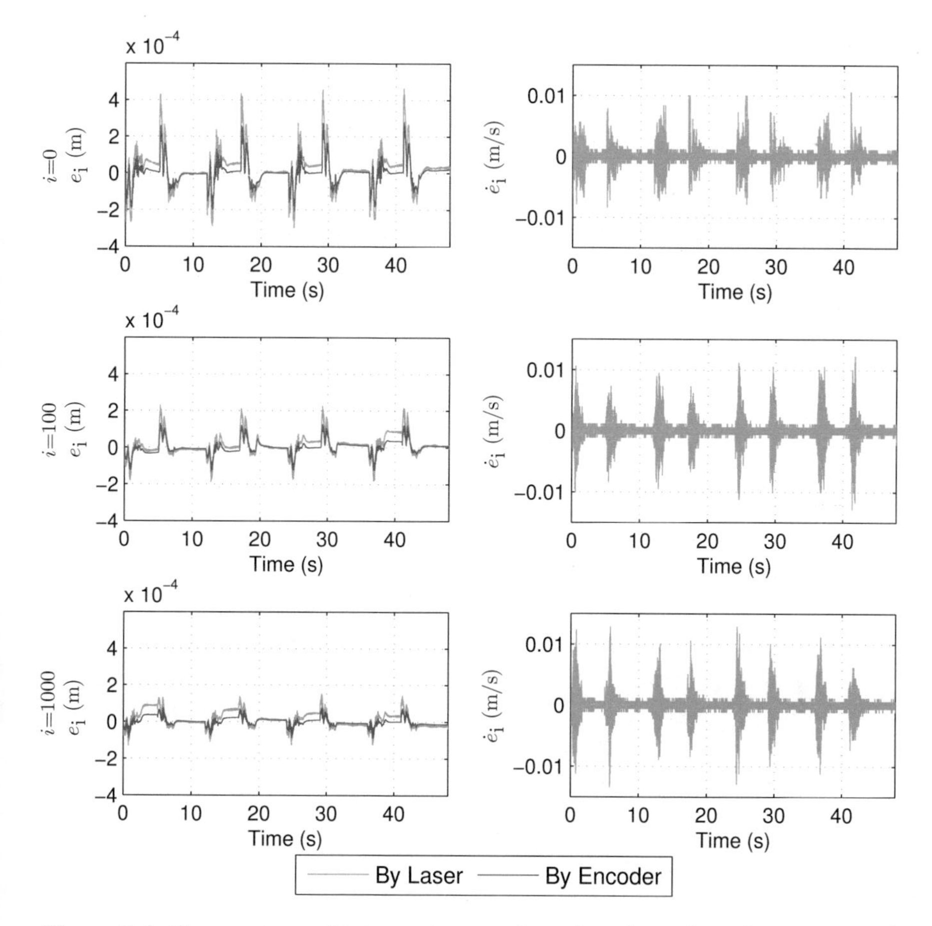

Figure 3.6: Comparison of inter-axis error based on laser interferometer and encoder with different iteration numbers. © [2017] Elsevier. Reprinted, with permission, from J. Ma, S.-L. Chen, N. Kamaldin, C. S. Teo, A. Tay, A. Al. Mamun, and K. K. Tan, A Novel Constrained H_2 Optimization Algorithm for Mechatronics Design in Flexure-Linked Biaxial Gantry, ISA Transactions, vol. 71, pp. 467–479, 2017.

sets of gains. This indicates that the resonant modes are not excited by these sets of feedback gains and the selected flexure piece. For easy visualization, the chattering of laser interferometer measurement is obtained by backward differentiation, as shown in Figure 3.6 as well. Control efforts u_1 and u_2 are demonstrated in Figure 3.7. The plots of numerically differentiated control efforts $\dot{u}_1$ and $\dot{u}_2$ derived post-experiment are illustrated in Figure 3.8. The coupling force v from the flexure calculated from (3.56) and its chattering are shown in Figure 3.9. Notice that $\dot{u}_1$, $\dot{u}_2$ and $\dot{v}$ are obtained by differentiation without the use of any low-pass filter. Thus, it can be concluded that the optimization is successful in terms of improvement of the tracking and inter-axis error, in the exchange of reasonable amounts of chattering in the control signals and jerk of the flexure force.

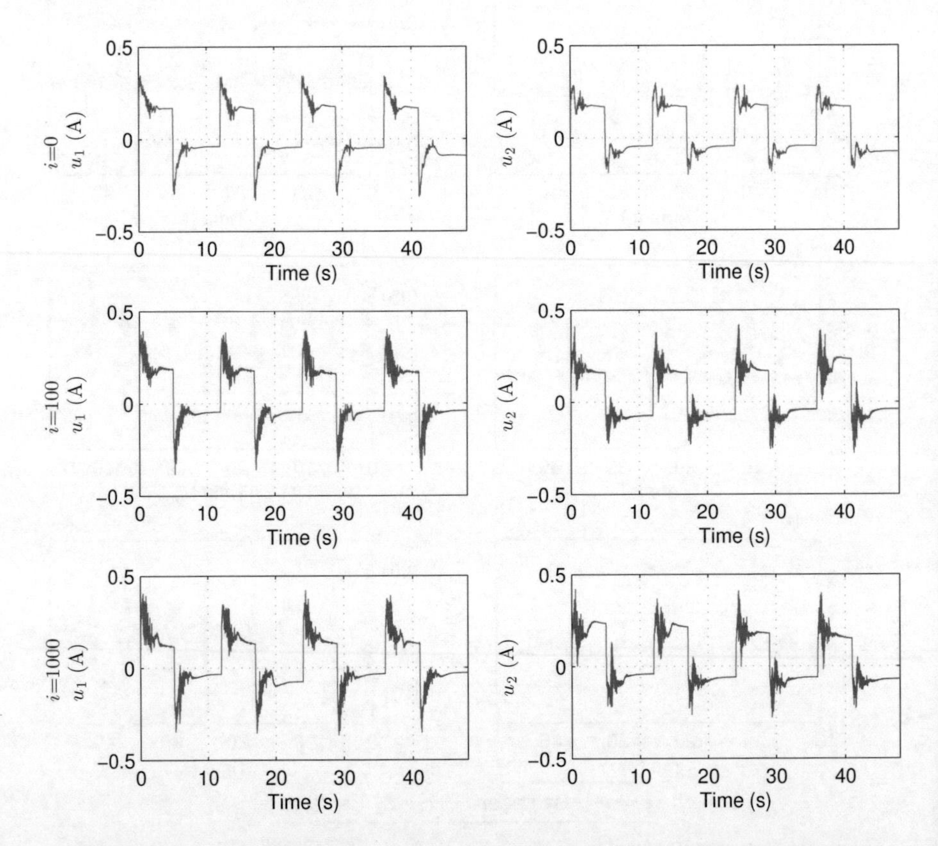

Figure 3.7: Comparison of control efforts with different iteration numbers. © [2017] Elsevier. Reprinted, with permission, from J. Ma, S.-L. Chen, N. Kamaldin, C. S. Teo, A. Tay, A. Al. Mamun, and K. K. Tan, A Novel Constrained H_2 Optimization Algorithm for Mechatronics Design in Flexure-Linked Biaxial Gantry, ISA Transactions, vol. 71, pp. 467–479, 2017.

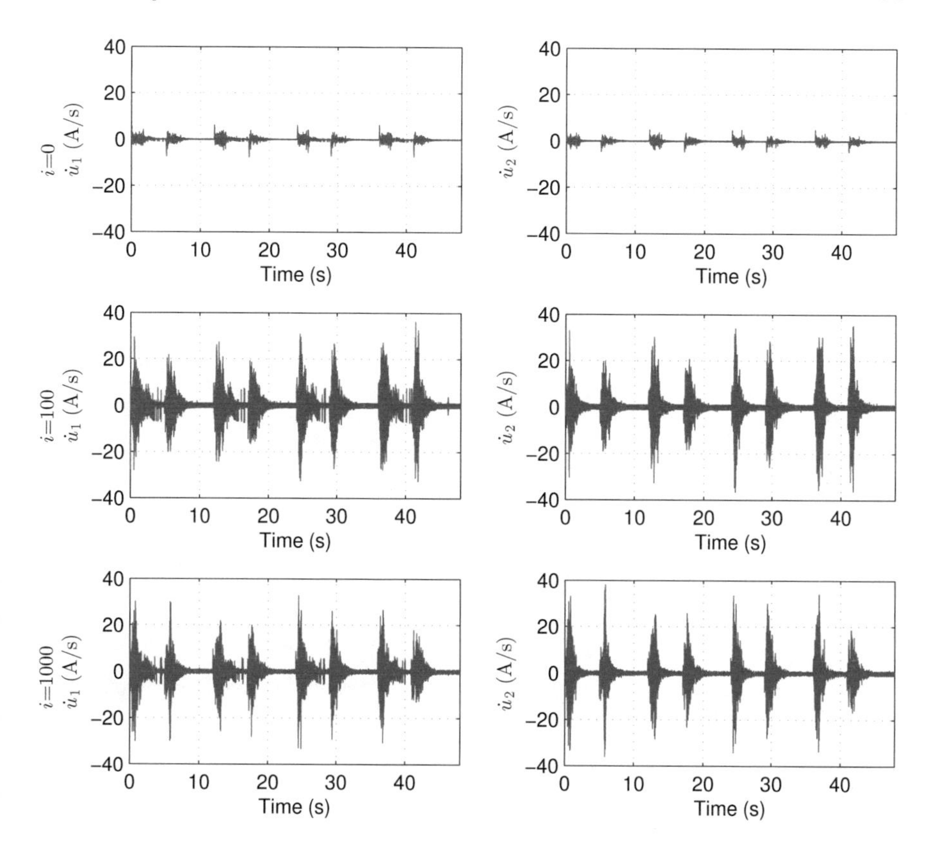

Figure 3.8: Comparison of chattering of control efforts with different iteration numbers. © [2017] Elsevier. Reprinted, with permission, from J. Ma, S.-L. Chen, N. Kamaldin, C. S. Teo, A. Tay, A. Al. Mamun, and K. K. Tan, A Novel Constrained H_2 Optimization Algorithm for Mechatronics Design in Flexure-Linked Biaxial Gantry, ISA Transactions, vol. 71, pp. 467–479, 2017.

The weightings should be defined by the user carefully. Overweighting on tracking performance will lead to obvious chattering in control input, flexure force, and position feedback, if the resonant modes of the end effector are excited. There are trade-offs among the different factors in the objective function. The weightings on these factors can be adjusted as long as the resonant modes of the end effector are not induced, which can be validated by the laser interferometer. Generally, optimized parameters by the proposed algorithm will guarantee a better performance according to the predefined cost function with a proper setting of weighting matrices.

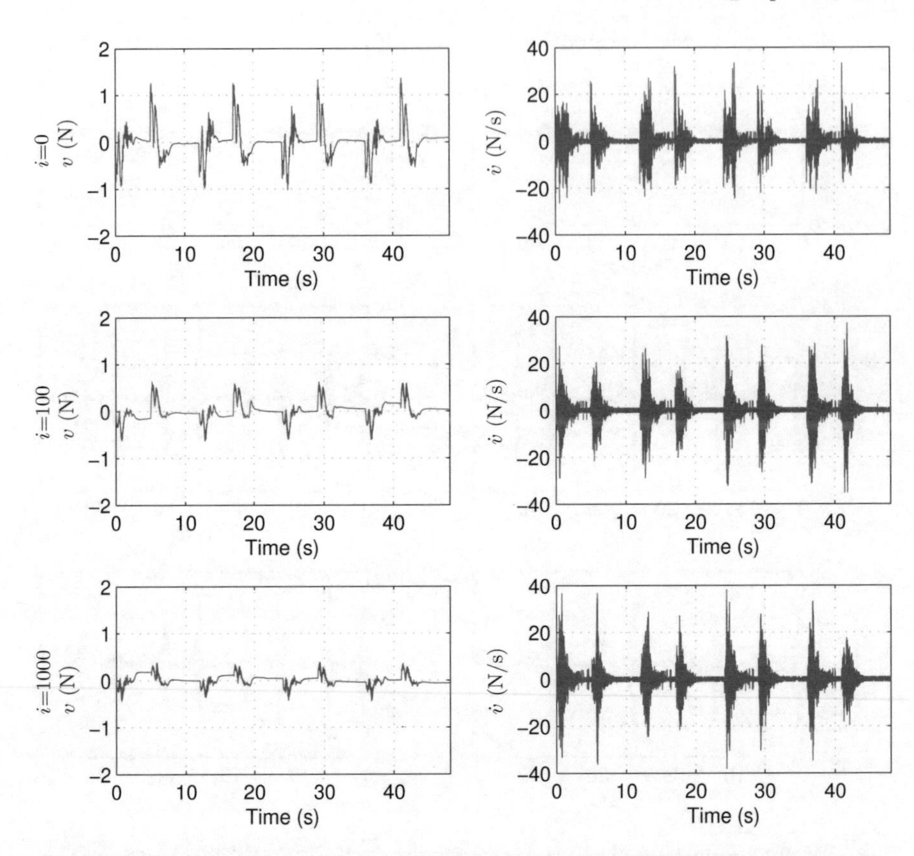

Figure 3.9: Comparison of flexure force and jerk with different iteration numbers. © [2017] Elsevier. Reprinted, with permission, from J. Ma, S.-L. Chen, N. Kamaldin, C. S. Teo, A. Tay, A. Al. Mamun, and K. K. Tan, A Novel Constrained H_2 Optimization Algorithm for Mechatronics Design in Flexure-Linked Biaxial Gantry, ISA Transactions, vol. 71, pp. 467–479, 2017.

3.3.4 Summary

In this work, it is aimed to improve the tracking performance of the DHG with flexure joints and minimize the possible induced chattering on the cross-arm, with respect to a class of point-to-point S-curve reference profiles of the same dynamics but different steps. A systematic integrated mechatronic design approach is developed, where the stiffness of the flexure and the parameters of decentralized PID controllers are optimized simultaneously. The design problem is quantified to an $\mathscr{H}_2$ optimization problem with penalization on the tracking error, the inter-axis error, the chattering of control inputs and the jerk of coupling force that resulted from the flexure joints, and the constraints due to the mechanical coupling force and the decentralized control structure are introduced. This mechatronic design problem is solved by the proposed constrained $\mathscr{H}_2$ optimization algorithm, and the effectiveness of the proposed method is validated by the experiments on the testbed.

3.4 Conclusion

As an extension of the constrained linear quadratic optimization algorithm, a constrained $\mathscr{H}_2$ optimization algorithm is proposed. The disturbance rejection is taken into consideration and the initial state dependency of the optimal solution is removed. This algorithm aims to optimize the required parameters by hybridizing direct computation of the projection gradient and line search of the optimal step iteratively, where the calculation of the unconstrained gradient for the standard $\mathscr{H}_2$ problem in both finite and infinite horizons, and the projection of the gradient onto the constrained hyperplane are summarized in respective theorems. Similar to the constrained linear quadratic optimization algorithm, by the execution of the proposed constrained $\mathscr{H}_2$ optimization algorithm, the cost of the objective function is monotonically decreasing while the stability of the closed-loop system is guaranteed throughout the optimization process. The algorithm can be used for a class of control system design problems as well. A case study on a mechatronic design problem in a flexure-linked DHG is addressed and the experimental results demonstrate the practical appeal of the proposed constrained $\mathscr{H}_2$ optimization algorithm.

4

Constrained $\mathscr{H}_2$ Guaranteed Cost Optimization

4.1 Background

In Chapter 2 and Chapter 3, projection gradient methods are introduced, and optimization algorithms are proposed to solve the constrained linear quadratic optimization problem and the constrained $\mathscr{H}_2$ optimization problem, but these algorithms can only find the local optimums and the model uncertainties are not explicitly handled. It is common that LQR and $\mathscr{H}_2$ control do not present guaranteed robustness a priori when model uncertainties exist. To overcome these limitations, the explicit account for model uncertainties using robust control design approach is necessary.

Although great efforts have been devoted in robust control for designing a control system which guarantees robust stability, it is desirable to design a control system which is not only stable but also guarantees an adequate level of performance. One approach to this problem is $\mathscr{H}_2$ guaranteed cost control, which aims to design a control system such that it guarantees an upper bound of quadratic performance for all admissible parameter uncertainties (Petersen and McFarlane 1994; Petersen 1995).

In fact, several controller design techniques are available to optimize the $\mathscr{H}_2$-norm of the closed-loop system with uncertain parameters (Geromel, Peres, and Souza 1992), and most of them are based on closed-loop system $\mathscr{H}_2$-norm expressions, or $\mathscr{H}_2$ guaranteed cost in the case of uncertain systems. For a precisely known open-loop system, the closed-loop system $\mathscr{H}_2$-norm can be easily determined (Doyle, Glover, Khargonekar, and Francis 1989; Palhares, Taicahashi, and Peres 1997). However, if parameter uncertainties exist in the open-loop system, it is impossible to calculate the $\mathscr{H}_2$-norm.

In (Petersen and Hollot 1986; Bernussou, Peres, and Geromel 1989; Barmish 1985), several approaches are presented for quadratic stabilizability issues in the case of uncertain linear systems, which aim to determine the stabilizing gain matrix for all feasible closed-loop models with uncertainties. Two major classes of structural uncertainties are considered, which are convex-bounded uncertainties (Geromel 1999; Peres, Geromel, and Bernussou 1993; Geromel, Peres, and Bernussou 1991) and norm-bounded uncertainties (Shi, Boukas, and Agarwal 1999; Han 2004; Zhou and Khargonekar

1988). Remarkably, convex-bounded uncertainties cover the important case of interval matrices (Geromel, Peres, and Souza 1992). Mixed $\mathscr{H}_2/\mathscr{H}_\infty$ control techniques proposed in (Khargonekar and Rotea 1991; Bernstein and Haddad 1989) are closely related to the $\mathscr{H}_2$ guaranteed cost control problem, but they cannot be applied when convex-bounded uncertainties exist.

Indeed, it is not an easy task to solve the $\mathscr{H}_2$ guaranteed cost control problem towards convex-bounded parameter uncertainties. On the other hand, certain motion control problems rely on the optimization of the gain matrix which is always subjected to structural constraints. Thus, it leaves an open problem to solve the $\mathscr{H}_2$ guaranteed cost control problem for uncertain systems with generalized feedback constraints.

4.2 Parameter Space Optimization with Structural Constraints

4.2.1 Problem Formulation

Consider an LTI system

$$\dot{x}(t) = \mathcal{A}x(t) + \mathcal{B}_2u(t) + \mathcal{B}_1w(t), \tag{4.1a}$$

$$z(t) = \mathcal{C}x(t) + \mathcal{D}u(t), \tag{4.1b}$$

with the control law

$$u(t) = -Kx(t), \tag{4.2}$$

where $x \in \mathbb{R}^n$ is the state vector with $x(0) = x_0$, $u \in \mathbb{R}^m$ is the control input vector, $w \in \mathbb{R}^l$ is the exogenous disturbance input, $z \in \mathbb{R}^q$ is the controlled output vector, $K \in \mathbb{R}^{m\times n}$ is the feedback gain matrix with structural constraints. Matrices $\mathcal{A} \in \mathbb{R}^{n\times n}$, $\mathcal{B}_2 \in \mathbb{R}^{n\times m}$, $\mathcal{B}_1 \in \mathbb{R}^{n\times l}$, $\mathcal{C} \in \mathbb{R}^{q\times n}$, $\mathcal{D} \in \mathbb{R}^{q\times m}$ are assumed to be known. Also, it is assumed that there is no cross weighting between the state variables and the control variables, i.e. $\mathcal{C}^T\mathcal{D} = 0$, the control weighting matrix is nonsingular, i.e. $\mathcal{D}^T\mathcal{D} > 0$, $(\mathcal{A}, \mathcal{B}_2)$ is stabilizable, the pair $(\mathcal{A}, \mathcal{C})$ has no unobservable modes on the imaginary axis.

The objective is to find K such that the objective function is minimized with an impulse function $w(t)$, where the objective function is defined as

$$J = \int_0^\infty z(t)^T z(t)\, dt. \tag{4.3}$$

It is equivalent to minimize the $\mathscr{H}_2$-norm of the transfer function

$$H(s) = (\mathcal{C} - \mathcal{D}K)(sI - \mathcal{A}_c)^{-1}\mathcal{B}_1 \tag{4.4}$$

from w to z, where $\mathcal{A}_c = \mathcal{A} - \mathcal{B}_2 K$. That is

$$\begin{aligned} J(K) &= \|H(s)\|_2^2 \\ &= \mathrm{Tr}((\mathcal{C} - \mathcal{D}K)W_c(\mathcal{C} - \mathcal{D}K)^T) \\ &= \mathrm{Tr}(\mathcal{B}_1^T W_o \mathcal{B}_1), \end{aligned} \tag{4.5}$$

where W_c and W_o are controllability and observability Gramians associated with the closed-loop system, satisfying

$$\mathcal{A}_c W_c + W_c \mathcal{A}_c^T + \mathcal{B}_1 \mathcal{B}_1^T = 0, \tag{4.6}$$

$$\mathcal{A}_c^T W_o + W_o \mathcal{A}_c + \mathcal{C}^T \mathcal{C} + K^T \mathcal{D}^T \mathcal{D} K = 0. \tag{4.7}$$

For systems with feedback structural constraints and model uncertainties, the gain K is not directly solvable by the standard formulae. Then, if model uncertainties exist in the system (4.1a), the above optimization problem can be generalized as an $\mathscr{H}_2$ guaranteed cost control problem with a constrained gain matrix.

4.2.2 Transformations of Feedback Gain Matrix with Generalized Structural Constraints

The strategy is to make transformations on K with the arbitrary structural constraints. To be more specific, it is assumed that K is non-decentralized with equal or opposite elements. First, the problem to optimize K is converted to optimize K_D with desirable structural properties (decentralized with equal or opposite elements) by state vector extension and system transformation, where a new state vector and a new system representation are constructed. This transformation can be expressed by $K = K_\mathrm{D} T$. Second, the problem to optimize K_D is transformed to optimize a decentralized unconstrained gain K_U, where $K_\mathrm{D} = K_\mathrm{U} H$. Notice that $K_\mathrm{D} = K_\mathrm{U} H$ is equivalent to $K_\mathrm{D} Y = 0$, with $Y = I - (H^T(HH^T)^{-1}H)$, i.e. $\mathscr{R}(K_\mathrm{D}) \subseteq \mathscr{R}(H)$, where $\mathscr{R}(\cdot)$ represents the range space of a matrix.

It is obvious that these transformations are factorizations of K into the product of K_D and T, and then K_D into the product of K_U and H. Indeed, there are infinite numbers of these factorizations. To complete such transformations, Algorithm 4.1 and Algorithm 4.2 are proposed based on Assumption 4.1.

Assumption 4.1 *K does not contain any zero column.*

Remark 4.1 *For each pair of equal or opposite elements, we need to remove either one in the pair and retain all the other elements to construct $\tilde{L}_i$. Two types of equality constraint are addressed in this work: equal elements (e.g. in matrix $\begin{bmatrix} a & a \end{bmatrix}$, "a" and "a" is a pair), and opposite elements (e.g. in matrix $\begin{bmatrix} a & -a \end{bmatrix}$, "a" and "-a" is a pair). To construct $\tilde{L}_i$, we need to retain only one element in the pair.*

Algorithm 4.1 Transformation from K to K_D

Require: $K \in \mathbb{R}^{m\times n}, N_i = 0, L_i = \emptyset, \forall i = 1, \ldots, m$
  for $i = 1 : m$ **do**
    for $j = 1 : n$ **do**
      if $K_{ij} \neq 0$ **then**
        $N_i \leftarrow N_i + 1$
        $L_i \leftarrow \text{append}(L_i, K_{ij})$
      end if
    end for
  end for
  return $K_\text{D} = \text{blocdiag}\{L_1, \ldots, L_m\}$

Algorithm 4.2 Transformation from K_D to K_U

  for $i = 1 : m$ **do**
    if $n_i = 0$ **then** $//n_i$ is the number of equal or opposite element pairs in the i-th row of K_D
      $\tilde{N}_i = N_i$
      $\tilde{L}_i = L_i$
    else
      $\tilde{N}_i = N_i - n_i$
      Construct $\tilde{L}_i$ by Remark 4.1
    end if
  end for
  return $K_\text{U} = \text{blocdiag}\{\tilde{L}_1, \ldots, \tilde{L}_m\}$

To conclude the above discussions, Theorem 4.1 is presented.

Theorem 4.1 *The following two statements hold.*

1. *For a given $K \in \mathbb{R}^{m\times n}$, it can be converted to $K_\text{D} = \text{blocdiag}\{L_1, \ldots, L_m\} \in \mathbb{R}^{m\times \bar{n}}$ by Algorithm 4.1, such that $u = -K_\text{D}x_\text{D}(t) = -Kx(t)$. Here, $L_i^T \in \mathbb{R}^{N_i}$, $\forall i = 1, \ldots, m$, $\bar{n} = \sum_{i=1}^m N_i$. There exists a transformation matrix $T \in \mathbb{R}^{\bar{n}\times n}$, such that $K = K_\text{D}T$ and $x_\text{D} = Tx \in \mathbb{R}^{\bar{n}}$ is a valid reconstructed state vector with $\bar{n} > n$.*

2. *When there are equal or opposite element constraints in K_D, there exists a transformation matrix $H \in \mathbb{R}^{\tilde{n}\times \bar{n}}$, such that $K_\text{D} = K_\text{U}H$, $K_\text{U} \in \mathbb{R}^{m\times \tilde{n}}$ with no constraint on K_U. The derivation of K_U is given by Algorithm 4.2, where $K_\text{U} = \text{blocdiag}\{\tilde{L}_1, \ldots, \tilde{L}_m\}$, and $\tilde{L}_i^T \in \mathbb{R}^{\tilde{N}_i}$, $\forall i = 1, \ldots, m$, $\tilde{n} = \sum_{i=1}^m \tilde{N}_i$, $\tilde{n} < \bar{n}$.*

Proof of Theorem 4.1: Theorem 4.1 is a direct consequence of Algorithm 4.1 and Algorithm 4.2.

Remark 4.2 *H is unique for given K_{D} and K_{U}, but T is not unique and it depends on the selection of x_{D}. The reconstructed system matrices are changed accordingly while the optimal solution is invariant with respect to different possible choices of T and x_{D}.*

Example 4.1 is given to illustrate the above results.

Example 4.1 *Given the state vector $x^T = \begin{bmatrix} x_1 & x_2 & x_3 \end{bmatrix}$ and the controller gain $K = \begin{bmatrix} a & b & 0 \\ 0 & c & c \end{bmatrix}$, it can be seen that K has non-decentralized structure with two equal elements in the second row. By Algorithm 4.1, K can be factorized as $K = K_{\mathrm{D}}T$, where $K_{\mathrm{D}} = \begin{bmatrix} a & b & 0 & 0 \\ 0 & 0 & c & c \end{bmatrix}$ is decentralized but still with equal element constraint, the new state vector can be defined as $x_{\mathrm{D}}^T = \begin{bmatrix} x_1 & x_2 & x_2 & x_3 \end{bmatrix}$, with the transformation matrix $T = \begin{bmatrix} 1 & 0 & 0 \\ 0 & 1 & 0 \\ 0 & 1 & 0 \\ 0 & 0 & 1 \end{bmatrix}$. Another possible factorization of K yields the same K_{D} but different $x_{\mathrm{D}}^T = \begin{bmatrix} x_1 & x_2 & x_3 & x_2 \end{bmatrix}$ and $T = \begin{bmatrix} 1 & 0 & 0 \\ 0 & 1 & 0 \\ 0 & 0 & 1 \\ 0 & 1 & 0 \end{bmatrix}$.*

Then, by Algorithm 4.2, we have $K_{\mathrm{U}} = \begin{bmatrix} a & b & 0 \\ 0 & 0 & c \end{bmatrix}$. It can be observed that K_{U} is decentralized without any additional constraint. Furthermore, we can obtain $H = \begin{bmatrix} 1 & 0 & 0 & 0 \\ 0 & 1 & 0 & 0 \\ 0 & 0 & 1 & 1 \end{bmatrix}$.

Notice that when K is transformed to K_{D}, the state-space model (4.1a)–(4.1b) with $\mathcal{A}$, $\mathcal{B}_2$, $\mathcal{B}_1$, $\mathcal{C}$, $\mathcal{D}$ is extended to the counterpart, where

$$\dot{x}_{\mathrm{D}}(t) = Ax_{\mathrm{D}}(t) + B_2u(t) + B_1w(t), \tag{4.8a}$$

$$z(t) = Cx_{\mathrm{D}}(t) + Du(t), \tag{4.8b}$$

with $A \in \mathbb{R}^{\bar{n}\times\bar{n}}$, $B_2 \in \mathbb{R}^{\bar{n}\times m}$, $B_1 \in \mathbb{R}^{\bar{n}\times l}$, $C \in \mathbb{R}^{q\times\bar{n}}$, $D \in \mathbb{R}^{q\times m}$. The reconstructed system matrices are given by $A = T\mathcal{A}(T^TT)^{-1}T^T$, $B_2 = T\mathcal{B}_2$, $B_1 = T\mathcal{B}_1$, $C = \mathcal{C}(T^TT)^{-1}T^T$, $D = \mathcal{D}$.

Example 4.2 is given to illustrate the reconstruction techniques on system matrices.

Example 4.2 *Suppose the original system is given by*

$$\begin{bmatrix} \dot{x}_1(t) \\ \dot{x}_2(t) \\ \dot{x}_3(t) \end{bmatrix} = \begin{bmatrix} a_{11} & a_{12} & a_{13} \\ a_{21} & a_{22} & a_{23} \\ a_{31} & a_{32} & a_{33} \end{bmatrix} \begin{bmatrix} x_1(t) \\ x_2(t) \\ x_3(t) \end{bmatrix} + \begin{bmatrix} b_{11} & b_{12} \\ b_{21} & b_{22} \\ b_{31} & b_{32} \end{bmatrix} \begin{bmatrix} u_1(t) \\ u_2(t) \end{bmatrix} + \begin{bmatrix} c_{11} \\ c_{21} \\ c_{31} \end{bmatrix} w(t),$$

and the first choice of T and x_{D} in the above example is selected for illustration purpose, where the extended state vector is assumed to be $x_{\mathrm{D}}^T = \begin{bmatrix} x_1 & x_2 & x_2 & x_3 \end{bmatrix}$, *with the transformation matrix* $T = \begin{bmatrix} 1 & 0 & 0 \\ 0 & 1 & 0 \\ 0 & 1 & 0 \\ 0 & 0 & 1 \end{bmatrix}$. *Then, the state-space representation for the reconstructed system is given by*

$$\begin{bmatrix} \dot{x}_1(t) \\ \dot{x}_2(t) \\ \dot{x}_2(t) \\ \dot{x}_3(t) \end{bmatrix} = \begin{bmatrix} a_{11} & a_{12} & 0 & a_{13} \\ a_{21} & a_{22} & 0 & a_{23} \\ a_{21} & a_{22} & 0 & a_{23} \\ a_{31} & a_{32} & 0 & a_{33} \end{bmatrix} \begin{bmatrix} x_1(t) \\ x_2(t) \\ x_2(t) \\ x_3(t) \end{bmatrix} + \begin{bmatrix} b_{11} & b_{12} \\ b_{21} & b_{22} \\ b_{21} & b_{22} \\ b_{31} & b_{32} \end{bmatrix} \begin{bmatrix} u_1(t) \\ u_2(t) \end{bmatrix} + \begin{bmatrix} c_{11} \\ c_{21} \\ c_{21} \\ c_{31} \end{bmatrix} w(t).$$

4.2.3 Setup of Feasible Sets for Generalized Feedback Constraints

To represent the open-loop model (4.8a)-(4.8b), the following extended matrices (Barmish 1983) are introduced, which exhibit important properties to be used in the sequel.

$$F = \begin{bmatrix} A & -B_2 \\ 0 & 0 \end{bmatrix} \in \mathbb{R}^{\bar{p}\times\bar{p}}, \quad G = \begin{bmatrix} 0 \\ I \end{bmatrix} \in \mathbb{R}^{\bar{p}\times m},$$

$$Q = \begin{bmatrix} B_1 B_1^T & 0 \\ 0 & 0 \end{bmatrix} \in \mathbb{R}^{\bar{p}\times\bar{p}}, \quad R = \begin{bmatrix} C^T C & 0 \\ 0 & D^T D \end{bmatrix} \in \mathbb{R}^{\bar{p}\times\bar{p}}, \tag{4.9}$$

where $\bar{p} = m + \bar{n}$.

Define the matrix

$$W = W^T = \begin{bmatrix} W_1 & W_2 \\ W_2^T & W_3 \end{bmatrix} \in \mathbb{R}^{\bar{p}\times\bar{p}}, \tag{4.10}$$

with $W_1 > 0 \in \mathbb{R}^{\bar{n}\times\bar{n}}$, $W_2 \in \mathbb{R}^{\bar{n}\times m}$, $W_3 \in \mathbb{R}^{m\times m}$ and define $\mathscr{N}$ as the null space of G^T except the origin, i.e. $\mathscr{N} = \{v \neq 0 \in \mathbb{R}^{\bar{p}} : G^T v = 0\}$. Hence, $v \in \mathscr{N}$ is of the form $v^T = \begin{bmatrix} x_{\mathrm{D}}^T & 0 \end{bmatrix}$ with $x_{\mathrm{D}} \neq 0$.

In addition, define the matrical function $\Theta(W) = FW + WF^T + Q$ and partition it as

$$\Theta(W) = \begin{bmatrix} \Theta_1(W) & \Theta_2(W) \\ \Theta_2^T(W) & \Theta_3(W) \end{bmatrix}, \tag{4.11}$$

with $\Theta_1(W) \in \mathbb{R}^{\bar{n}\times\bar{n}}, \Theta_2(W) \in \mathbb{R}^{\bar{n}\times m}, \Theta_3(W) \in \mathbb{R}^{m\times m}$. Then, the following theorem bridges a feasible set about W to the closed-loop stability of the system.

Theorem 4.2 *Define the set* $\mathscr{C}_1 = \{W : W = W^T \geq 0, \Theta_1(W) \leq 0\}$. *Then the following statements hold.*

1. $\mathscr{C}_1 \neq \emptyset$ *if and only if* (A, B_2) *is stabilizable.*

2. $\mathscr{C}_1$ *is a convex set.*

Proof of Theorem 4.2: For Statement 1, if (A, B_2) is stabilizable, there exists $W_\mathrm{p} \geq W_\mathrm{c} > 0$ such that

$$(A - B_2 K_\mathrm{F})W_\mathrm{p} + W_\mathrm{p}(A - B_2 K_\mathrm{F})^T + B_1 B_1^T \leq 0, \tag{4.12}$$

where K_F is a gain matrix with no prescribed structural constraint. Meanwhile, from $\Theta(W) = FW + WF^T + Q$, we have

$$\Theta_1(W) = AW_1 - B_2 W_2^T + W_1 A^T - W_2 B_2^T + B_1 B_1^T. \tag{4.13}$$

Thus, from (4.12) and (4.13), by setting $W_1 = W_\mathrm{p}$ and $W_2^T = K_\mathrm{F} W_\mathrm{p}$, then we have $K_\mathrm{F} = W_2^T W_1^{-1}$ and $\Theta_1(W) \leq 0$. In addition, we can construct $W = \begin{bmatrix} W_1 & W_1 K_\mathrm{F}^T \\ K_\mathrm{F} W_1 & W_3 \end{bmatrix}$. By Schur complement, we can ensure $W \geq 0$ by choosing $W_3 \geq K_\mathrm{F} W_1 K_\mathrm{F}^T$. Hence the set $\mathscr{C}_1 \neq \emptyset$. The necessity can be proved similarly. First, it is worth mentioning that the set $\{W : \Theta_1(W) \leq 0\}$ is equivalent to the set $\{W : v^T \Theta(W) v \leq 0, \forall v \in \mathscr{N}\}$. For any symmetric positive semi-definite W partitioned as in (4.10) with $W_1 > 0$, we have

$$\begin{aligned} v^T\Theta(W)v &= x^T\Theta_1(W)x \\ &= x^T(AW_1 - B_2W_2^T + W_1A^T - W_2B_2^T + B_1B_1^T)x \\ &= x^T((A - B_2W_2^TW_1^{-1})W_1 + W_1(A - B_2W_2^TW_1^{-1})^T + B_1B_1^T)x \\ &\leq 0. \end{aligned} \tag{4.14}$$

Therefore, the gain $K_\mathrm{F} = W_2^T W_1^{-1}$ is stabilizing for the pair (A, B_2).

For Statement 2, the set of all positive semi-definite W is a convex cone, and $\Theta(W)$ is affine with respect to W. It is easy to see $\mathscr{C}_1$ is the intersection of infinite numbers of open half spaces. This concludes the proof of Theorem 4.2.

$W \in \mathscr{C}_1$ is a sufficient but not necessary condition to stabilize the closed-loop system. However, $W \geq 0$ is reserved to provide a norm bound of the gain K_F, since $W_3 \geq K_\mathrm{F} W_1 K_\mathrm{F}^T$ provided $W_1 > 0$ (Geromel, Peres, and Bernussou 1991), where $K_\mathrm{F} = W_2^T W_1{}^{-1}$ is a stabilizing gain for pair (A, B_2). It is important to know that K_F is a full matrix without any imposed structural constraint and W has no structural constraint as well in this scenario. However, in the case that the gain matrix is under a specific decentralized structure as K_D, the following property provides a guideline to constrain the structure of W, such that the specific decentralized structure of K_D is preserved.

Property 4.1 *To let K_D preserve the specific decentralized structure*

$$K_\mathrm{D} = \mathrm{blocdiag}\{K_{\mathrm{D},1}, \ldots, K_{\mathrm{D},m}\}, \tag{4.15}$$

with $K_{\mathrm{D},i}^T \in \mathbb{R}^{\mathrm{D}_i}$, $\forall i = 1, \ldots, m$, one can define $W \in \mathscr{C}_{1\mathrm{D}} = \mathscr{C}_1 \cap \mathscr{C}_\mathrm{D}$, where $\mathscr{C}_\mathrm{D} = \{W : W_1 = W_{1\mathrm{D}}, W_2 = W_{2\mathrm{D}}\}$, and

$$W_{1\mathrm{D}} = \mathrm{blocdiag}\{W_{1\mathrm{D},1}, \ldots, W_{1\mathrm{D},m}\}, \tag{4.16}$$
$$W_{2\mathrm{D}} = \mathrm{blocdiag}\{W_{2\mathrm{D},1}, \ldots, W_{2\mathrm{D},m}\}, \tag{4.17}$$

with $W_{1\mathrm{D,i}} \in \mathbb{R}^{\mathrm{D}_i \times \mathrm{D}_i}$ *and* $W_{2\mathrm{D,i}} \in \mathbb{R}^{\mathrm{D}_i}$, $\forall i = 1, \ldots, m$.

Proof of Property 4.1: Property 4.1 can be directly validated by the given structural constraints. In other words, W_2^T has the same decentralized structure as K_D.

Definition 4.1 *The system* (4.8a) *is called decentralized stabilizable if there exists* $W \in \mathscr{C}_{1\mathrm{D}}$.

Theorem 4.3 *The set of all decentralized stabilizing gains* $\mathscr{K}_\mathrm{D} = \{K_\mathrm{D} = W_2^T W_1{}^{-1} : W \in \mathscr{C}_{1\mathrm{D}}\}$.

Proof of Theorem 4.3: The proof of Theorem 4.3 follows from the proof of Theorem 4.2 and Definition 4.1.

Before we establish a one-to-one mapping between W and a feasible K_U, we define a real-valued function $f(W)$ that exhibits important properties as summarized by the following theorem (Geromel, Peres, and Souza 1993).

Theorem 4.4 *Define the function* $f(\cdot) : domf \to \mathbb{R}$, *where*

$$f(W) = \mathrm{Tr}(W_2^T W_1^{-1} W_2 - W_2^T H^T (H W_1 H^T)^{-1} H W_2), \tag{4.18}$$

with $domf = \{W \in \mathbb{R}^{\bar{p} \times \bar{p}} : W_1 > 0\}$. *If* $f(W) = 0$, *then* $\exists W \in domf$ *s.t.* $W_2^T W_1^{-1} Y = 0$, *and* $K_\mathrm{U} = W_2^T H^T (H W_1 H^T)^{-1}$. *With* $K_\mathrm{U} = W_2^T H^T (H W_1 H^T)^{-1}$ *and* $K_\mathrm{D} = W_2^T W_1^{-1}$,

$$\frac{df(W)}{dW} = \begin{bmatrix} H^T K_\mathrm{U}^T K_\mathrm{U} H - K_\mathrm{D}^T K_\mathrm{D} & K_\mathrm{D}^T - H^T K_\mathrm{U}^T \\ K_\mathrm{D} - K_\mathrm{U} H & 0 \end{bmatrix}, \tag{4.19}$$

$$f(W) = \left\langle \frac{df(W)}{dW}, W \right\rangle = \left\| (W_2^T W_1^{-1} - W_2^T H^T (H W_1 H^T)^{-1} H) W_1^{1/2} \right\|^2, \tag{4.20}$$

$\forall W \in domf$, *where* $\langle A, B \rangle = \mathrm{Tr}(A^T B)$.

Proof of Theorem 4.4: Part of the proof is shown in (Geromel, Peres, and Souza 1993). To obtain $df(W)/dW$, Lemma 2.1 is used. In addition, the following lemma is introduced.

Lemma 4.1 *Suppose* $A \in \mathbb{R}^{n \times n}$, $B \in \mathbb{R}^{n \times n}$, *both* A *and* $(A + B)$ *are invertible, then* $(A + B)^{-1} = A^{-1} - A^{-1} B A^{-1} + O((BA^{-1})^2)$.

Define $W_+ = W + \varepsilon \Delta W$, $W_{i+} = W_i + \varepsilon \Delta W_i$, $\forall i = 1, 2$. From Lemma 2.1, it is easy to see

$$W_{1+}^{-1} = W_1^{-1} - \varepsilon W_1^{-1} \Delta W_1 W_1^{-1} + O(\varepsilon^2 (\Delta W_1 W_1^{-1})^2). \tag{4.21}$$

Similarly, we have

$$(HW_{1+}H^T)^{-1} = (HW_1H^T)^{-1} - \varepsilon (HW_1H^T)^{-1}(H\Delta W_1 H^T)(HW_1H^T)^{-1} \\ + O(\varepsilon^2((H\Delta W_1 H^T)(HW_1H^T)^{-1})^2). \tag{4.22}$$

Define $\Delta W = \begin{bmatrix} \Delta W_1 & \Delta W_2 \\ \Delta W_2^T & \Delta W_3 \end{bmatrix} \in \mathbb{R}^{\bar{p} \times \bar{p}}$, where $\Delta W_1 \in \mathbb{R}^{\bar{n} \times \bar{n}}$, $\Delta W_2 \in \mathbb{R}^{\bar{n} \times m}$ and $\Delta W_3 = 0 \in \mathbb{R}^{m \times m}$, we have

$$f(W_+) = \mathrm{Tr}(W_{2+}^T W_{1+}^{-1} W_{2+}) - \mathrm{Tr}(W_{2+}^T H^T (HW_{1+}H^T)^{-1} HW_{2+}), \tag{4.23}$$

with

$$W_{2+}^T W_{1+}^{-1} W_{2+} = W_2^T W_1^{-1} W_2 + \varepsilon (K_{\mathrm{D}} \Delta W_2 - K_{\mathrm{D}} \Delta W_1 K_{\mathrm{D}}^T + \Delta W_2^T K_{\mathrm{D}}^T), \tag{4.24}$$

and

$$W_{2+}^T H^T (HW_{1+}H^T)^{-1} HW_{2+} = W_2^T H^T (HW_1H^T)^{-1} HW_2 + \\ \varepsilon (K_{\mathrm{U}} H \Delta W_2 - K_{\mathrm{U}} H \Delta W_1 H^T K_{\mathrm{U}}^T + \Delta W_2^T H^T K_{\mathrm{U}}^T). \tag{4.25}$$

Thus, $f(W_+) = f(W) + \varepsilon \mathrm{Tr}(M(K_{\mathrm{D}}, K_{\mathrm{U}}, H)\Delta W)$, with

$$M(K_{\mathrm{D}}, K_{\mathrm{U}}, H) = \begin{bmatrix} H^T K_{\mathrm{U}}^T K_{\mathrm{U}} H - K_{\mathrm{D}}^T K_{\mathrm{D}} & K_{\mathrm{D}}^T - H^T K_{\mathrm{U}}^T \\ K_{\mathrm{D}} - K_{\mathrm{U}} H & 0 \end{bmatrix}. \tag{4.26}$$

Therefore, from Lemma 2.1, we have

$$\frac{df(W)}{dW} = M(K_{\mathrm{D}}, K_{\mathrm{U}}, H)^T = M(K_{\mathrm{D}}, K_{\mathrm{U}}, H). \tag{4.27}$$

Then, it is straightforward to complete the proof of Theorem 4.4.

From Theorem 4.4, $f(W) \geq 0$ for all $W \in \mathrm{dom} f$. Since the set that consists of all the stabilizing K_{D} is non-convex, it is important to stress that the projection condition $K_{\mathrm{D}} Y = 0$ is a non-convex constraint in the parameter space. Even though $f(W)$ is not convex, $f(W)$ reaches its global minimum $f(W) = 0$ when $df(W)/dW = 0$, or equivalently, $K_{\mathrm{D}} = K_{\mathrm{U}} H$. In other words, the following statements are equivalent: (a). $df(W)/dW = 0$; (b). $f(W) = 0$ with $W_1 > 0$; (c). $K_{\mathrm{D}} = K_{\mathrm{U}} H$. In this way, the factorization $K_{\mathrm{D}} = K_{\mathrm{U}} H$ can be interpreted in terms of the above-mentioned function, i.e. $f(W) = 0$.

Now, we eventually bridge W to the feedback gain K_{U} in the presence of equal or opposite element constraints, according to the following definition and theorem. Prior to that, we define $\mathscr{C}_2 = \{W : f(W) = 0\}$, and $\mathscr{C} = \mathscr{C}_{1\mathrm{D}} \cap \mathscr{C}_2$.

Definition 4.2 *The system* (4.8a) *is called constraint-restricted decentralized stabilizable if there exists* $W \in \mathscr{C}$.

Theorem 4.5 *Define* $\mathscr{K}_{\mathrm{U}} = \{K_{\mathrm{U}} = W_2^T H^T (HW_1H^T)^{-1} : W \in \mathscr{C}\}$, *then,*

1. *Any* $K_{\mathrm{U}} \in \mathscr{K}_{\mathrm{U}}$ *stabilizes the closed-loop system.*
2. *Any* $W \in \mathscr{C}$ *generates* $K_{\mathrm{U}} \in \mathscr{K}_{\mathrm{U}}$ *such that* $\mathrm{Tr}(RW) \geq \|H(s)\|_2^2$.
3. *At optimality,* $W^* = \mathrm{argmin}\{\mathrm{Tr}(RW) : W \in \mathscr{C}\}$ *yields* $K_{\mathrm{U}}^* \in \mathscr{K}_{\mathrm{U}}$, *s.t.* $J^* = \min \|H(s)\|_2^2 = \mathrm{Tr}(RW^*) \leq \mathrm{Tr}(RW)$.

Proof of Theorem 4.5: As in the proof of Theorem 4.2, $W_1 \geq W_{\mathrm{c}}$, then we have $W_3 \geq W_2^T W_1^{-1} W_2 = K_{\mathrm{U}} H W_1 (K_{\mathrm{U}} H)^T \geq K_{\mathrm{U}} H W_{\mathrm{c}} (K_{\mathrm{U}} H)^T$. So

$$\begin{aligned}\mathrm{Tr}(RW) &= \mathrm{Tr}(CW_1C^T + DW_3D^T) \\ &\geq \mathrm{Tr}(CW_{\mathrm{c}}C^T + DK_{\mathrm{U}}HW_{\mathrm{c}}(K_{\mathrm{U}}H)^T D^T) \\ &= \mathrm{Tr}((C - DK_{\mathrm{U}}H)W_{\mathrm{c}}(C - DK_{\mathrm{U}}H)^T) \\ &= \|H(s)\|_2^2. \end{aligned} \tag{4.28}$$

This concludes the proof of Theorem 4.5.

It implies that $\mathrm{Tr}(RW)$ provides an upper bound of $\|H(s)\|_2^2$. When the optimal K_{U}^* is found, the equality in Theorem 4.5(c) holds, while the $\mathscr{H}_2$-norm is minimized.

4.2.4 Extension to Uncertain Systems

We would like to extend the previous results to the uncertain systems, which can be represented by the convex combination of $(A_i, B_{2i}, B_{1i}, C, D)$, $\forall i = 1, \ldots, N$. Then, we make the following assumption.

Assumption 4.2 *The model uncertainties are structural and convex-bounded.*

Thus, $F = \sum_{i=1}^N \xi_i F_i$ belongs to a polyhedral domain, which can be expressed as a convex combination of the extreme matrices F_i, where $F_i = \begin{bmatrix} A_i & -B_{2i} \\ 0 & 0 \end{bmatrix} \in \mathbb{R}^{\bar{p} \times \bar{p}}$ represents the extreme vertex of the uncertain domain, $\xi_i \geq 0, \forall i = 1, \ldots, N$, and $\sum_{i=1}^N \xi_i = 1$. Remarkably, a precisely known system is a special case of the above expression, where $N = 1$.

In advance, define the matrical function $\Theta_i(W) = F_iW + WF_i^T + Q_i, \forall i = 1, \ldots, N$, where $Q_i = \begin{bmatrix} B_{1i}B_{1i}^T & 0 \\ 0 & 0 \end{bmatrix} \in \mathbb{R}^{\bar{p} \times \bar{p}}$, and $\Theta_i(W) = \begin{bmatrix} \Theta_{1i}(W) & \Theta_{2i}(W) \\ \Theta_{2i}^T(W) & \Theta_{3i}(W) \end{bmatrix}$, with $\Theta_{1i}(W) \in \mathbb{R}^{\bar{n} \times \bar{n}}, \Theta_{2i}(W) \in \mathbb{R}^{\bar{n} \times m}, \Theta_{3i}(W) \in \mathbb{R}^{m \times m}$. Now, we are able to construct a mapping between W and K_{U} by the following definition and theorem. Similarly, define $\mathscr{C}_1' = \bigcap_{i=1}^{N} \mathscr{C}_{1\mathrm{D}}^i$, where $\mathscr{C}_{1\mathrm{D}}^i = \{W : W = W^T \geq 0, \Theta_{1i}(W) \leq 0\} \cap \mathscr{C}_{\mathrm{D}}$, then $\mathscr{C}' = \mathscr{C}_1' \cap \mathscr{C}_2$.

Definition 4.3 *The system (4.8a) is called robustly constraint-restricted decentralized stabilizable if there exists $W \in \mathscr{C}'$.*

Theorem 4.6 *Define $\mathscr{K}_{\mathrm{U}}' = \{K_{\mathrm{U}} = W_2^T H^T (H W_1 H^T)^{-1} : W \in \mathscr{C}'\}$, then,*

1. *Any $K_{\mathrm{U}} \in \mathscr{K}_{\mathrm{U}}'$ stabilizes the closed-loop system in presence of the convex-bounded uncertainties.*
2. *Any $W \in \mathscr{C}'$ generates $K_{\mathrm{U}} \in \mathscr{K}_{\mathrm{U}}'$ such that $\mathrm{Tr}(RW) \geq \|H(s)\|_2^2$.*
3. *At optimality, $W^* = \mathrm{argmin}\{\mathrm{Tr}(RW) : W \in \mathscr{C}'\}$ yields $K_{\mathrm{U}}^* \in \mathscr{K}_{\mathrm{U}}'$, s.t. $J^* = \min \|H(s)\|_2^2 = \mathrm{Tr}(RW^*) \leq \mathrm{Tr}(RW)$.*

Proof of Theorem 4.6: The proof is straightforward since this theorem is an extension from Theorem 4.2 and Theorem 4.5.

The closed-loop stability for the overall domain needs to be checked at vertices of the convex polyhedron only. Therefore, if the number of uncertain parameters is N_u, then the stability for the entire uncertain domain is only required to be checked at $N = 2^{N_u}$ vertices.

4.3 Constrained $\mathscr{H}_2$ Guaranteed Cost Optimization Algorithm

The matrix W that admits the optimal gain matrix lies in $\mathscr{C}'$, which is the intersection of convex set $\mathscr{C}_1'$ and non-convex set $\mathscr{C}_2$, but $\mathscr{C}'$ is not explicitly known. To overcome the difficulty, linear programming is conducted within a pre-defined polytope $\mathscr{P}^l$. If W^l is located outside $\mathscr{C}'$, cutting plane technique is used and then a half space $\mathcal{S}^l$ is constructed for separation purposes and update of the polytope. Utilizing "Optimization-Checking-Cutting" strategy, the following numerical procedures are proposed such that the global solution for the optimization problem is found in an iterative framework.

Theorem 4.7 *For the case when the most violated constraint is in $\mathscr{C}_1'$:*

1. *If W^l violates the constraint (c1), i.e. $\lambda_{\mathrm{m}}(W^l) < 0$, then the half space containing $\mathscr{C}_1'$ but not containing W^l can be defined as*

$$\mathcal{S}^l = \{W : \langle v_0 v_0^T, W \rangle \geq 0\}, \tag{4.29}$$

where v_0 is the unit-norm eigenvector corresponding to $\lambda_{\mathrm{m}}(W^l)$.

2. If W^l violates the constraint (c2), i.e. $\lambda_{\mathrm{M}}(\Theta_{1i}(W^l)) > 0$, the half space containing $\mathscr{C}_1'$ but not containing W^l can be defined as

$$\mathcal{S}^l = \{W : \langle v_{0i}v_{0i}^T F_i + F_i^T v_{0i}v_{0i}^T, W\rangle \leq -v_{0i}^T Q_i v_{0i}\}, \tag{4.30}$$

where $v_{0i}^T = \begin{bmatrix} x_{0i}^T & 0 \end{bmatrix}$, $v_{0i} \in \mathbb{R}^{\bar{p}}$, and $x_{0i} \in \mathbb{R}^{\bar{n}}$ is the unit-norm eigenvector corresponding to $\lambda_{\mathrm{M}}(\Theta_{1i}(W^l))$.

Algorithm 4.3 Constrained $\mathscr{H}_2$ Guaranteed Cost Optimization Algorithm

---

- Step 1: Initialize an arbitrary small $\varepsilon > 0$. Set the iteration index $l = 0$ and define a polytope $\mathcal{P}^0 = \{W : W_1 = W_{1\mathrm{D}}, W_2 = W_{2\mathrm{D}}, W = W^T, W_{jj} \geq 0, \forall j = 1, \ldots, \bar{p}\}$.

- Step 2: Set $l = l + 1$, and solve the linear programming problem: $W^l = \operatorname{argmin}\{\mathrm{Tr}(RW) : W \in \mathcal{P}^l\}$ and $J^l = \mathrm{Tr}(RW^l)$.

- Step 3: ε-optimality test.

  This step is to check whether constraints (c1), (c2) and (c3) are satisfied after W^l is obtained. Here, the maximum and the minimum eigenvalues of a matrix are denoted by $\lambda_{\mathrm{M}}(\cdot)$ and $\lambda_{\mathrm{m}}(\cdot)$, respectively.

  3.1 If (c1) is violated, calculate $\lambda_{\mathrm{m}}(W^l)$. Otherwise, set $\lambda_{\mathrm{m}}(W^l) = 0$;

  3.2 If (c2) is violated, calculate $\lambda_{\mathrm{M}}(\Theta_{1i}(W^l))$. Otherwise, set $\lambda_{\mathrm{M}}(\Theta_{1i}(W^l)) = 0$;

  3.3 If (c3) is violated, assign the value of $f(W^l)$ to $\lambda_{\mathrm{f}}(W^l)$. Otherwise, set $\lambda_{\mathrm{f}}(W^l) = 0$.

  3.4 If $|\lambda_{\mathrm{m}}(W^l)| \leq \varepsilon$, $|\lambda_{\mathrm{M}}(\Theta_{1i}(W^l))| \leq \varepsilon$, $\forall i = 1, \ldots, N$, and $|\lambda_{\mathrm{f}}(W^l)| \leq \varepsilon$, calculate $K_{\mathrm{U}} = W_2^T H^T (HW_1H^T)^{-1}$, and the upper bound of $\min \|H(s)\|_2^2$ is estimated as $J^* = J^l$, then go to Step 4. Otherwise, go to Step 5.

- Step 4: Calculate K from K_{U} and verify the closed-loop stability over the polyhedral domain of model uncertainties. If the stability is ensured, the optimization is done; otherwise, go back to Step 1 and decrease the ε.

- Step 5: Define the most violated constraint as the one corresponding to $\max\{|\lambda_{\mathrm{m}}(W^l)|, |\lambda_{\mathrm{M}}(\Theta_{1i}(W^l))|\,\forall i = 1, \ldots, N, |\lambda_{\mathrm{f}}(W^l)|\}$.

- Step 6: Generate the cutting plane according to the most violated constraint determined in Step 5.

  6.1 If the most violated constraint is in $\mathscr{C}_1'$, generate a cutting plane $\mathcal{S}^l$ where $\mathscr{C}_1' \subseteq \mathcal{S}^l$ such that $W^l \notin \mathcal{S}^l$, based on Theorem 4.7.

  6.2 If the most violated constraint is in $\mathscr{C}_2$, i.e. $f(W^l) > 0$, generate the linear subspace $\mathcal{S}^l$ as the separating hyperplane based on Theorem 4.8.

- Step 7: Update the polytope as $\mathcal{P}^{l+1} = \mathcal{P}^l \cap \mathcal{S}^l$, then go back to Step 2.

Proof of Theorem 4.7: From $\langle v_0 v_0^T, W\rangle \geq 0$, we have $\mathrm{Tr}(v_0^T W v_0) \geq 0$, which implies $v_0^T W v_0 \geq 0$. Obviously (4.29) can be used as a cutting plane for separation if W^l violates the constraint (c1). From $\langle v_{0i} v_{0i}^T F_i + F_i^T v_{0i} v_{0i}^T, W\rangle \leq -v_{0i}^T Q_i v_{0i}$, we have $v_{0i}^T W F_i^T v_{0i} + v_{0i}^T F_i W v_{0i} \leq -v_{0i}^T Q_i v_{0i}$, or $v_{0i}^T \Theta_i(W) v_{0i} \leq 0$. This is equivalent to $x_{0i}^T \Theta_{1i}(W) x_{0i} \leq 0$.

Similarly, (4.30) can be used as a cutting plane for separation if W^l violates the constraint (c2). This concludes the proof of Theorem 4.7.

Theorem 4.8 *For the case when the most violated constraint is in $\mathscr{C}_2$: If W^l violates the constraint (c3), i.e. $f(W^l) > 0$, then the half space containing all W such that $f(W) \leq 0$ but not containing W^l can be defined as*

$$\mathcal{S}^l = \left\{ W : \left\langle \frac{df(W^l)}{dW^l}, W \right\rangle \leq 0 \right\}. \tag{4.31}$$

Proof of Theorem 4.8: Define the conjugate function of $f(W)$ (Boyd and Vandenberghe 2004): $f^\star(W^\star) = \sup\{\langle W, W^\star\rangle - f(W)\}$; there exists a domain $\chi = \{W : f^\star(df(W)/dW) < \infty\} \subseteq \mathrm{dom} f$, such that $\forall W^l \in \chi$ and $\forall W \in \mathrm{dom} f$, we have

$$\begin{aligned} f(W) &\geq f(W^l) + \left\langle \frac{df(W^l)}{dW^l}, W - W^l \right\rangle \\ &= \left\langle \frac{df(W^l)}{dW^l}, W^l \right\rangle + \left\langle \frac{df(W^l)}{dW^l}, W - W^l \right\rangle \\ &= \left\langle \frac{df(W^l)}{dW^l}, W \right\rangle. \end{aligned} \tag{4.32}$$

For $W^\star = df(W)/dW$, we can determine the supporting hyperplane to the epi$f$ at any $W \in \mathrm{dom} f$. We assume that during all iterations, $W \in \chi \subseteq \mathrm{dom} f$. Thus, it is obvious that if $f(W^l) > 0$, the half space $\mathcal{S}^l = \{W : \langle df(W^l)/dW^l, W\rangle \leq 0\}$ does not contain W^l and contains all W such that $f(W) \leq 0$, i.e. $f(W) = 0$. Due to the geometrical properties of the function $f(\cdot)$, the cutting plane technique can be successfully applied, and the global solution is attained within the domain χ. More details can be found in (Peres, Geromel, and Souza 1993). This concludes the proof of Theorem 4.8.

In Step 1, to achieve $W \in \mathscr{C}'$, it requires five types of constraints on W: (c1). $W \geq 0$; (c2). $\Theta_{1i}(W) \leq 0$, $\forall i = 1, \ldots, N$; (c3). $f(W) = 0$; (c4). $W_1 = W_{1\mathrm{D}}, W_2 = W_{2\mathrm{D}}$; (c5). $W = W^T$. We choose (c4) and (c5), rather than (c1), (c2) and (c3), to define the initial polytope, because the former ones can be incorporated in the linear programming routine. In addition, to guarantee the existence of an initial feasible solution for $\min \mathrm{Tr}(RW)$, additional $W_{jj} \geq 0, \forall j = 1, \ldots, \bar{p}$ are included in the linear programming. In this way, $\mathscr{C}' \subseteq \mathcal{P}^0$.

It is worth mentioning that Step 7 can be done by introducing an additional constraint $W^{l+1} \in \mathcal{S}^l$ to the linear programming in the last iteration, because $W^{l+1} \in \mathcal{P}^{l+1}$ leads to $W^{l+1} \in \mathcal{P}^{l}$ and $W^{l+1} \in \mathcal{S}^{l}$. The cutting planes are implemented as linear inequality constraints in the linear programming routine. Before reaching the stopping criterion, one inequality constraint is incorporated in each iteration, so the objective function is monotonically increasing through iterations. The cost at the final iteration is the minimum achievable value given that all other constraints are satisfied.

The above cutting-plane method is actually an outer linearization technique. This technique estimates an unknown convex set by sequentially adding a series of linear constraints, without causing infeasibility (Bertsekas and Yu 2011). Eventually, the estimated set contains and closely fits the targeted set by arbitrary small "distance" of ε.

4.4 Case Study

For simplification, "(t)" is omitted in the expression of time-history signals in this section.

4.4.1 Statement of Problem

We take the air bearing DHG as an example in this section. It is known that maintaining good tracking and synchronization performance of two carriages via either controller or mechanical design is necessary. However, for parallel mechanisms such as DHGs, it is hard to perform accurate system identification over the entire workspace, due to the strong inter-axial coupling force (Sharifzadeh, Arian, Salimi, Masouleh, and Kalhor 2017; Chen, Wu, Deng, and Wang 2017; Castro-Garcia, Tiels, Agudelo, and Suykens 2018). Hence, the motion controller should be synthesized to cater to such model uncertainties (Hao, Wang, Zhao, and Wang 2016; Chevalier, Copot, Ionescu, and De Keyser 2016). Similar to Chapter 2 and Chapter 3, the axes in the DHG are controlled by independent PID controllers, and such a decentralized control scheme imposes additional constraints for optimal controller synthesis.

In this work, associated symbols for all parameters are shown in Table 4.1. The schematic diagrams of the DHG are illustrated in Figure 3.1(a) and Figure 3.1(b), and the DHG is modeled as a coupled linear system with parametric uncertainties, where

$$(M_1 + \Delta M_1)\ddot{y}_1 = K_{\mathrm{f}} u_1 - \Gamma_1 \dot{y}_1 + v - f_1 w, \tag{4.33a}$$

$$(M_2 + \Delta M_2)\ddot{y}_2 = K_{\mathrm{f}} u_2 - \Gamma_2 \dot{y}_2 - v - f_2 w, \tag{4.33b}$$

with $v = k_{\mathrm{v}}(y_2 - y_1)$, $M_1 = (m_{\mathrm{e}} + m_{\mathrm{c}})/2 + m_1$ and $M_2 = (m_{\mathrm{e}} + m_{\mathrm{c}})/2 + m_2$, and $w(t)$ is a unit step function. Here, we assume that the damping coefficients of two axes are the same, where $\Gamma_1 = \Gamma_2 = \Gamma$.

Table 4.1: Nomenclature used in Chapter 4

Name	Unit	Description
K_{f}	N/A	Force constant
m_1	kg	Mass of carriage 1
m_2	kg	Mass of carriage 2
m_{e}	kg	Mass of end-effector
m_{c}	kg	Mass of cross-arm
Γ_1	Ns/m	Damping coefficient in carriage 1
Γ_2	Ns/m	Damping coefficient in carriage 2
f_1	N	Coulomb friction in carriage 1
f_2	N	Coulomb friction in carriage 2
y_1	m	Position of carriage 1
y_2	m	Position of carriage 2
y_{d}	m	Reference position for two carriages
Θ	rad	Yaw angle between two carriages
u_1	A	Control current of carriage 1
u_2	A	Control current of carriage 2
$u_{1\mathrm{ff}}$	A	Feedforward control current of carriage 1
$u_{2\mathrm{ff}}$	A	Feedforward control current of carriage 2
$u_{1\mathrm{fb}}$	A	Feedback control current of carriage 1
$u_{2\mathrm{fb}}$	A	Feedback control current of carriage 2
v	N	Coupling force from flexure
k_{v}	N/m	Stiffness of flexure

An S-curve trajectory $y_{\mathrm{d}} = p_1$ is used as the reference profile, where $\dot{p} = A_{\mathrm{p}}p$, with $p^T = \begin{bmatrix} p_1 & p_2 & p_3 \end{bmatrix}$, $A_{\mathrm{p}} = \begin{bmatrix} 0 & 1 & 0 \\ 0 & 0 & 1 \\ a_{\mathrm{p}1} & a_{\mathrm{p}2} & a_{\mathrm{p}3} \end{bmatrix}$. It is aimed to keep the precision tracking of two axes with minimizing induced chattering from feedback control signals.

4.4.2 Formulation of Constrained $\mathscr{H}_2$ Guaranteed Cost Optimization Problem

The controlled output vector z is defined as

$$z = C_{\mathrm{z}} \begin{bmatrix} e_1 & e_2 & \dot{u}_{1\mathrm{fb}} & \dot{u}_{2\mathrm{fb}} \end{bmatrix}^T, \tag{4.34}$$

where $C_{\mathrm{z}} = \mathrm{diag}\{q_1, q_2, r_1, r_2\}$, $e_1 = y_{\mathrm{d}} - y_1$ and $e_2 = y_{\mathrm{d}} - y_2$. Then, the objective function for coming optimization is defined as

$$J = \int_0^{\infty} z^T z \, dt. \tag{4.35}$$

To be tallied with common industrial control architecture, in this work, a 2-DOF control scheme composed of feedforward terms $u_{i\mathrm{ff}}$ and feedback terms $u_{i\mathrm{fb}}$ is employed, where

$$u_1 = u_{1\mathrm{ff}} + u_{1\mathrm{fb}}, \tag{4.36a}$$
$$u_2 = u_{2\mathrm{ff}} + u_{2\mathrm{fb}}. \tag{4.36b}$$

Due to the limited information exchange between two carriages, the decentralized control structure is proposed. Here, based on the nominal model of the DHG, the decentralized feedforward controller is designed in advance as

$$u_{1\mathrm{ff}} = \frac{\Gamma_1}{K_{\mathrm{f}}}\dot{y}_{\mathrm{d}} + \frac{M_1}{K_{\mathrm{f}}}\ddot{y}_{\mathrm{d}}, \tag{4.37a}$$
$$u_{2\mathrm{ff}} = \frac{\Gamma_2}{K_{\mathrm{f}}}\dot{y}_{\mathrm{d}} + \frac{M_2}{K_{\mathrm{f}}}\ddot{y}_{\mathrm{d}}. \tag{4.37b}$$

Define the matrix perturbed by parametric uncertainties as $\widetilde{(\bullet)} = (\bullet) + \Delta(\bullet)$; (4.33) is converted to the error dynamics

$$\widetilde{M}_1\ddot{e}_1 = \Delta M_1\ddot{y}_{\mathrm{d}} - K_{\mathrm{f}}u_{1\mathrm{fb}} - \Gamma_1\dot{e}_1 - v + f_1 w, \tag{4.38a}$$
$$\widetilde{M}_2\ddot{e}_2 = \Delta M_2\ddot{y}_{\mathrm{d}} - K_{\mathrm{f}}u_{2\mathrm{fb}} - \Gamma_2\dot{e}_2 + v + f_2 w, \tag{4.38b}$$

with $v = k_{\mathrm{v}}(e_1 - e_2)$.

Define $x^T = \begin{bmatrix} e_1 & \dot{e}_1 & \ddot{e}_1 & e_2 & \dot{e}_2 & \ddot{e}_2 & p_1 & p_2 & p_3 \end{bmatrix}$ and $u_{\mathrm{fb}}^T = \begin{bmatrix} u_{1\mathrm{fb}} & u_{2\mathrm{fb}} \end{bmatrix}$; the state-space representation of the augmented system (4.38) is given by

$$\dot{x} = \widetilde{A}x + \widetilde{B}_2\dot{u}_{\mathrm{fb}} + \widetilde{B}_1\dot{w}, \tag{4.39}$$

with

$$A = \begin{bmatrix} A_{\mathrm{s}} & 0_{6\times 3} \\ 0_{3\times 6} & A_{\mathrm{p}} \end{bmatrix}, \quad B_2^T = \begin{bmatrix} B_{2\mathrm{s}}^T & 0_{2\times 3} \end{bmatrix}, \quad B_1^T = \begin{bmatrix} B_{1\mathrm{s}}^T & 0_{1\times 3} \end{bmatrix},$$
$$\Delta A = \begin{bmatrix} \Delta A_{\mathrm{s}} & \Delta A_{\mathrm{s}}' \\ 0_{3\times 6} & 0_{3\times 3} \end{bmatrix}, \quad \Delta B_2^T = \begin{bmatrix} \Delta B_{2\mathrm{s}}^T & 0_{2\times 3} \end{bmatrix}, \quad \Delta B_1^T = \begin{bmatrix} \Delta B_{1\mathrm{s}}^T & 0_{1\times 3} \end{bmatrix}. \tag{4.40}$$

Define $M_{1\Delta} = \Delta M_1/(M_1 + \Delta M_1)$, $M_{2\Delta} = \Delta M_2/(M_2 + \Delta M_2)$, and $\hat{M}_{1\Delta} = M_{1\Delta}/M_1$, $\hat{M}_{2\Delta} = M_{2\Delta}/M_2$; various matrices in the state-space model (4.40) are defined as

$$A_{\mathrm{s}} = \begin{bmatrix} 0 & 1 & 0 & 0 & 0 & 0 \\ 0 & 0 & 1 & 0 & 0 & 0 \\ 0 & -\frac{K_{\mathrm{v}}}{M_1} & -\frac{\Gamma_1}{M_1} & 0 & \frac{K_{\mathrm{v}}}{M_1} & 0 \\ 0 & 0 & 0 & 0 & 1 & 0 \\ 0 & 0 & 0 & 0 & 0 & 1 \\ 0 & \frac{K_{\mathrm{v}}}{M_2} & 0 & 0 & -\frac{K_{\mathrm{v}}}{M_2} & -\frac{\Gamma_2}{M_2} \end{bmatrix},$$

$$\Delta A_{\mathrm{s}} = \begin{bmatrix} 0 & 0 & 0 & 0 & 0 & 0 \\ 0 & 0 & 0 & 0 & 0 & 0 \\ 0 & K_{\mathrm{v}}\hat{M}_{1\Delta} & \Gamma_1\hat{M}_{1\Delta} & 0 & -K_{\mathrm{v}}\hat{M}_{1\Delta} & 0 \\ 0 & 0 & 0 & 0 & 0 & 0 \\ 0 & 0 & 0 & 0 & 0 & 0 \\ 0 & -K_{\mathrm{v}}\hat{M}_{2\Delta} & 0 & 0 & K_{\mathrm{v}}\hat{M}_{2\Delta} & \Gamma_2\hat{M}_{2\Delta} \end{bmatrix},$$

$$\Delta A'_{\mathrm{s}} = \begin{bmatrix} 0 & 0 & 0 \\ 0 & 0 & 0 \\ a_{\mathrm{p1}}M_{1\Delta} & a_{\mathrm{p2}}M_{1\Delta} & a_{\mathrm{p3}}M_{1\Delta} \\ 0 & 0 & 0 \\ 0 & 0 & 0 \\ a_{\mathrm{p1}}M_{2\Delta} & a_{\mathrm{p2}}M_{2\Delta} & a_{\mathrm{p3}}M_{2\Delta} \end{bmatrix},$$

$$B_2 = \begin{bmatrix} 0 & 0 & -\frac{K_{\mathrm{f}}}{M_1} & 0 & 0 & 0 & 0 & 0 & 0 \\ 0 & 0 & 0 & 0 & 0 & -\frac{K_{\mathrm{f}}}{M_2} & 0 & 0 & 0 \end{bmatrix}^T,$$

$$\Delta B_2 = \begin{bmatrix} 0 & 0 & K_{\mathrm{f}}\hat{M}_{1\Delta} & 0 & 0 & 0 & 0 & 0 & 0 \\ 0 & 0 & 0 & 0 & 0 & K_{\mathrm{f}}\hat{M}_{2\Delta} & 0 & 0 & 0 \end{bmatrix}^T,$$

$$B_1 = \begin{bmatrix} 0 & 0 & \frac{f_1}{M_1} & 0 & 0 & \frac{f_2}{M_2} & 0 & 0 & 0 \end{bmatrix}^T,$$

$$\Delta B_1 = \begin{bmatrix} 0 & 0 & -f_1\hat{M}_{1\Delta} & 0 & 0 & -f_2\hat{M}_{2\Delta} & 0 & 0 & 0 \end{bmatrix}^T. \tag{4.41}$$

The feedback gain is denoted by K; from $\dot{u_{\mathrm{fb}}} = -Kx$, the feedback controller is obtained as

$$u_{\mathrm{fb}} = -K\int_0^t x\,d\tau. \tag{4.42}$$

To preserve the constraint from the decentralized nature of PID control architecture, the feedback gain is given by

$$K = -\mathrm{blocdiag}\{K_1, K_2\}, \tag{4.43}$$

with

$$K_1 = \begin{bmatrix} k_{\mathrm{i1}} & k_{\mathrm{p1}} & k_{\mathrm{d1}} \end{bmatrix}, \tag{4.44}$$

$$K_2 = \begin{bmatrix} k_{\mathrm{i2}} & k_{\mathrm{p2}} & k_{\mathrm{d2}} & 0 & 0 & 0 \end{bmatrix}. \tag{4.45}$$

Notice that K_2 contains additional 3 zero elements, due to the augmentation of 3 states of the S-curve trajectory. These states are uncontrollable by the decentralized controller, but they are stabilizable autonomously.

Hence, with given K, a pair of decentralized PID controllers is eventually obtained as

$$u_{1\text{fb}} = k_{\text{p}1}e_1 + k_{\text{i}1}\int_0^t e_1\,d\tau + k_{\text{d}1}\dot{e}_1, \tag{4.46a}$$

$$u_{2\text{fb}} = k_{\text{p}2}e_2 + k_{\text{i}2}\int_0^t e_2\,d\tau + k_{\text{d}2}\dot{e}_2. \tag{4.46b}$$

Meanwhile, the controlled output z in (4.34) is expressed in the following equivalent form as

$$z = Cx + D\dot{u}_{\text{fb}}, \tag{4.47}$$

with $C = Q_{\text{C}} \cdot \begin{bmatrix} \begin{bmatrix} 1 & 0 & 0 & 0 & 0 & 0 & 0 & 0 & 0 \\ 0 & 0 & 0 & 1 & 0 & 0 & 0 & 0 & 0 \end{bmatrix} \\ 0_{2\times 9} \end{bmatrix} \in \mathbb{R}^{4\times 9}$, $D = R_{\text{D}} \cdot \begin{bmatrix} 0_{2\times 2} \\ I_2 \end{bmatrix} \in \mathbb{R}^{4\times 2}$, where the weighting matrices are defined as $Q_{\text{C}} = \text{diag}\{q_1, q_2, 0, 0\}$, $R_{\text{D}} = \text{diag}\{0, 0, r_1, r_2\}$. Here, the transfer function from $\dot{w}$ to z is given by

$$H(s) = C_{\text{c}}(sI - A_{\text{c}})^{-1}B_1, \tag{4.48}$$

where $A_{\text{c}} = A - B_2K$, $C_{\text{c}} = C - DK$. Since $\dot{w}$ is an impulse function, the optimization problem (4.35) is equivalent to minimize the $\mathscr{H}_2$-norm of $H(s)$. That is

$$\begin{aligned} J &= \|H(s)\|_2^2 \\ &= \text{Tr}\big(C_{\text{c}}W_{\text{c}}C_{\text{c}}^T\big) \\ &= \text{Tr}(B_1^T W_{\text{o}} B_1), \end{aligned} \tag{4.49}$$

where W_{c} and W_{o} are controllability and observability Gramians associated with the closed-loop system satisfying the following two equations

$$A_{\text{c}}W_{\text{c}} + W_{\text{c}}A_{\text{c}}^T + B_1B_1^T = 0, \tag{4.50}$$

$$A_{\text{c}}^T W_{\text{o}} + W_{\text{o}}A_{\text{c}} + C_{\text{c}}^T C_{\text{c}} = 0. \tag{4.51}$$

To this end, together with (4.37), the 2-DOF decentralized control is to be designed for the system, following the optimization criteria (4.35), where the block diagram is shown in Figure 4.1.

If the system is accurately modeled and there is no constraint on the feedback gain K, it is well known that K is directly solvable by the AREs. However, in our case, the system model is perturbed by parametric uncertainties as in (4.38). In addition, from (4.43)-(4.45), the feedback gain K is restricted to the decentralized form, and the closed-loop system has some uncontrollable,

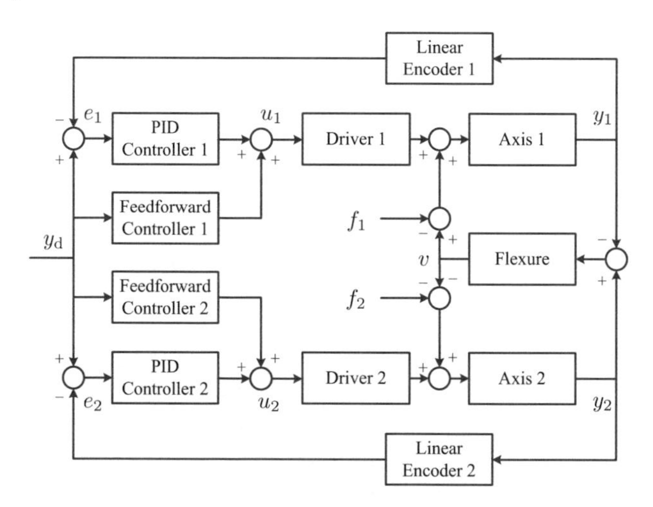

Figure 4.1: Overview of decentralized control architecture. © [2019] IEEE. Reprinted, with permission, from J. Ma, S.-L. Chen, W. Liang, C. S. Teo, A. Tay, A. Al. Mamun, and K. K. Tan, Robust Decentralized Controller Synthesis in Flexure-Linked H-Gantry by Iterative Linear Programming, IEEE Transactions on Industrial Informatics, vol. 15, no. 3, pp. 1698–1708, 2019.

but stabilizable states. In this case, no direct solution from the AREs is available. Hence, it is a challenging problem to synthesize K effectively under such feedback structural constraints and model uncertainties, when the method of solving the AREs is no longer applicable. Thus, the proposed cutting plane based optimization algorithm can be used.

4.4.3 Optimization and Experimental Validation

As stated in Chapter 2 and Chapter 3, it is common that for the flexure-linked DHG, only a series of flexure pieces with different stiffness is available for replacement. To find the optimal controller gains, we aim to minimize the objective function (4.35) for DHG equipped with finite choices of flexure pieces. In this work, we use the 2-mm and the 3-mm flexure joints as examples and we need to choose the most suitable one between them. Such a problem is generalized as the robust decentralized controller design problem for the uncertain system. Thus, constraint (c3) in Step 3 of Algorithm 4.3 is always satisfied, then we have $H = I$. Since K already has the decentralized structure, then $K_{\mathrm{U}} = K_{\mathrm{D}} = K = W_2 W_1^{-1}$. In other words, this mechatronic design problem for the flexure-linked DHG serves as a special case by using Algorithm 4.3.

Theorem 4.1 can only preserve the decentralized structure of the gain matrix, but additional constraints exist in (4.45) due to certain zero elements, so we have the following corollary.

Corollary 4.1 *To preserve the following special decentralized structure of K, where*

$$K = \text{blocdiag}\{K_1, \ldots, K_m\}, \tag{4.52}$$

$$K_m = \begin{bmatrix} K_{m,1} & K_{m,2} \end{bmatrix}, \tag{4.53}$$

with $K_i^T \in \mathbb{R}^{n_i}$, $\forall i = 1, \ldots, m$ with $n_1 + \ldots + n_m = n$, and $K_{m,1}^T \in \mathbb{R}^{n_{m,1}}$, $K_{m,2} = 0_{1\times n_{m,2}}$ and $n_{m,1} + n_{m,2} = n_m$, one can redefine

$$W \in \mathscr{C}_{\text{D}}, \tag{4.54}$$

where $\mathscr{C}_{\text{D}} = \{W : W_1 = W_{1\text{D}}, W_{1,m} = W_{1\text{D},m}, W_2 = W_{2\text{D}}, W_{2,m} = W_{2\text{D},m}\}$, and $W_{1\text{D}} = \text{blocdiag}\{W_{1\text{D},1}, \ldots, W_{1\text{D},m}\}$, $W_{2\text{D}} = \text{blocdiag}\{W_{2\text{D},1}, \ldots, W_{2\text{D},m}\}$, $W_{1\text{D},m} = \text{blocdiag}\{W_{1\text{D},m,1}, W_{1\text{D},\text{m},2}\}$, $W_{2\text{D},m}^T = \begin{bmatrix} W_{2\text{D},m,1}^T & W_{2\text{D},m,2}^T \end{bmatrix}$, with $W_{1\text{D},i} \in \mathbb{R}^{n_i \times n_i}$ and $W_{2\text{D},i} \in \mathbb{R}^{n_i}$, $\forall i = 1, \ldots, m$, $W_{1\text{D},m,1} \in \mathbb{R}^{n_{m,1}\times n_{m,1}}$, $W_{1\text{D},m,2} \in \mathbb{R}^{n_{m,2}\times n_{m,2}}$, $W_{2\text{D},m,1} \in \mathbb{R}^{n_{m,1}}$, and $W_{2\text{D},m,2} = 0 \in \mathbb{R}^{n_{m,2}}$.

Proof of Corollary 4.1: As an extension of Theorem 4.1, the proof is straightforward.

The model parameters remain the same as in Chapter 3. The reference profile in Figure 3.3 is used, the stopping criterion is defined as $\varepsilon = 10^{-6}$, and the weightings C_z in (4.34) are set to $q_1 = q_2 = 1$, $r_1 = r_2 = 1$. We assume the uncertainties ΔM_1 and ΔM_2 are within $\pm 10\%$ range of M_1 and M_2. When the 2-mm flexure is used, the stopping criterion is met after 608 iterations, W_1, W_2 and W_3 are obtained as

$$W_1 = 10^{-8} \times \text{blocdiag}\left\{ \begin{bmatrix} 0 & -1.2 & 57.7 \\ -1.2 & 1.1 & -12.2 \\ 57.7 & -12.2 & 1052.2 \end{bmatrix}, \right.$$
$$\left. \begin{bmatrix} 138.1 & 0.7 & -34.4 \\ 0.7 & 0.8 & 0.7 \\ -34.4 & 0.7 & 713.2 \end{bmatrix}, \begin{bmatrix} 88.6 & -111.7 & 13.6 \\ -111.7 & 146.0 & -17.0 \\ 13.6 & -17.0 & 56.3 \end{bmatrix} \right\}, \tag{4.55}$$

$$W_2^T = 10^{-8} \times \text{blocdiag}\{ \begin{bmatrix} 193.0 & -194.4 & 363.5 \end{bmatrix},$$
$$\begin{bmatrix} -194.3 & -151.1 & -363.6 & 0 & 0 & 0 \end{bmatrix} \}, \tag{4.56}$$

$$W_3 = 10^{-8} \times \begin{bmatrix} 814.6 & 0 \\ 0 & 780.7 \end{bmatrix}. \tag{4.57}$$

From $K = W_2^T W_1^{-1}$, the optimal controller is given as

$$K = -\text{blocdiag}\{ \begin{bmatrix} 18.5 & 216.5 & 1.2 \end{bmatrix},$$
$$\begin{bmatrix} 0.5 & 196.3 & 0.3 & 0 & 0 & 0 \end{bmatrix} \}, \tag{4.58}$$

and the estimation of upper bound to $\|H(s)\|_2^2$ is given by 1.73×10^{-5}. It can be seen that both W and K preserve the specific decentralized structure, which is coincident with Corollary 4.1.

Next, we use the 3-mm flexure to do the optimization. After 361 iterations, W_1, W_2 and W_3 are obtained as

$$W_1 = 10^{-8} \times \text{blocdiag}\left\{ \begin{bmatrix} 0 & -0.1 & 12.6 \\ -0.1 & 0.1 & -4.7 \\ 12.6 & -4.7 & 643.6 \end{bmatrix}, \right.$$

$$\left. \begin{bmatrix} 0 & 0.2 & -15.8 \\ 0.2 & 0.3 & -6.9 \\ -15.8 & -6.9 & 853.2 \end{bmatrix}, \begin{bmatrix} 77.2 & -146.9 & 117.0 \\ -146.9 & 297.0 & -303.0 \\ 117.0 & -303.0 & 612.6 \end{bmatrix} \right\}, \tag{4.59}$$

$$\begin{aligned} W_2^T = 10^{-8} \times \text{blocdiag}\{ & \begin{bmatrix} 32.5 & -136.6 & 152.9 \end{bmatrix}, \\ & \begin{bmatrix} -39.5 & -232.2 & 9\ .5 & 0 & 0 & 0 \end{bmatrix} \}, \end{aligned} \tag{4.60}$$

$$W_3 = 10^{-8} \times \begin{bmatrix} 269.8 & 89.6 \\ 89.6 & 639.2 \end{bmatrix}. \tag{4.61}$$

Thus, the optimal controller is given by

$$\begin{aligned} K = -\text{blocdiag}\{ & \begin{bmatrix} 126.2 & 2340.0 & 1\ .2 \end{bmatrix}, \\ & \begin{bmatrix} 91.0 & 909.8 & 9.0 & 0 & 0 & 0 \end{bmatrix} \}, \end{aligned} \tag{4.62}$$

and the upper bound to $\|H(s)\|_2^2$ is given by 9.09×10^{-6}.

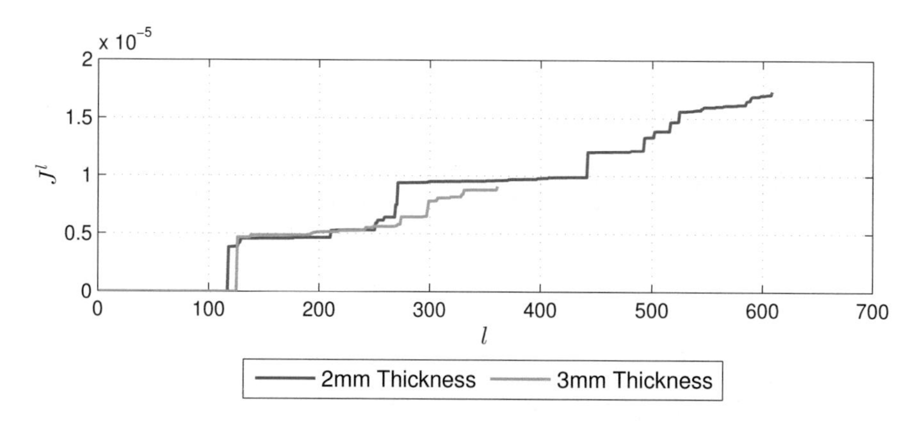

Figure 4.2: Evolution of J^l. © [2019] IEEE. Reprinted, with permission, from J. Ma, S.-L. Chen, W. Liang, C. S. Teo, A. Tay, A. Al. Mamun, and K. K. Tan, Robust Decentralized Controller Synthesis in Flexure-Linked H-Gantry by Iterative Linear Programming, IEEE Transactions on Industrial Informatics, vol. 15, no. 3, pp. 1698–1708, 2019.

The evolutions of J^l when selecting the 2-mm and the 3-mm flexures are illustrated in Figure 4.2, where $J^l = \text{Tr}(RW^l)$. It can be seen that with the addition of linear constraints in each iteration, $\text{Tr}(RW^l)$ is increasing until the stopping criterions are met. Comparing the 2-mm and the 3-mm flexure joints, the latter one results in a tighter upper bound to $\|H(s)\|_2^2$. So we choose the 3-mm flexure for the following experiments.

In the coming test, the experiments are conducted using the 3-mm flexure with optimized controller gains. Tracking errors, control efforts and their chattering of two axes are illustrated in Figure 4.3. Second, to validate the robustness of optimal parameters obtained using the proposed optimization

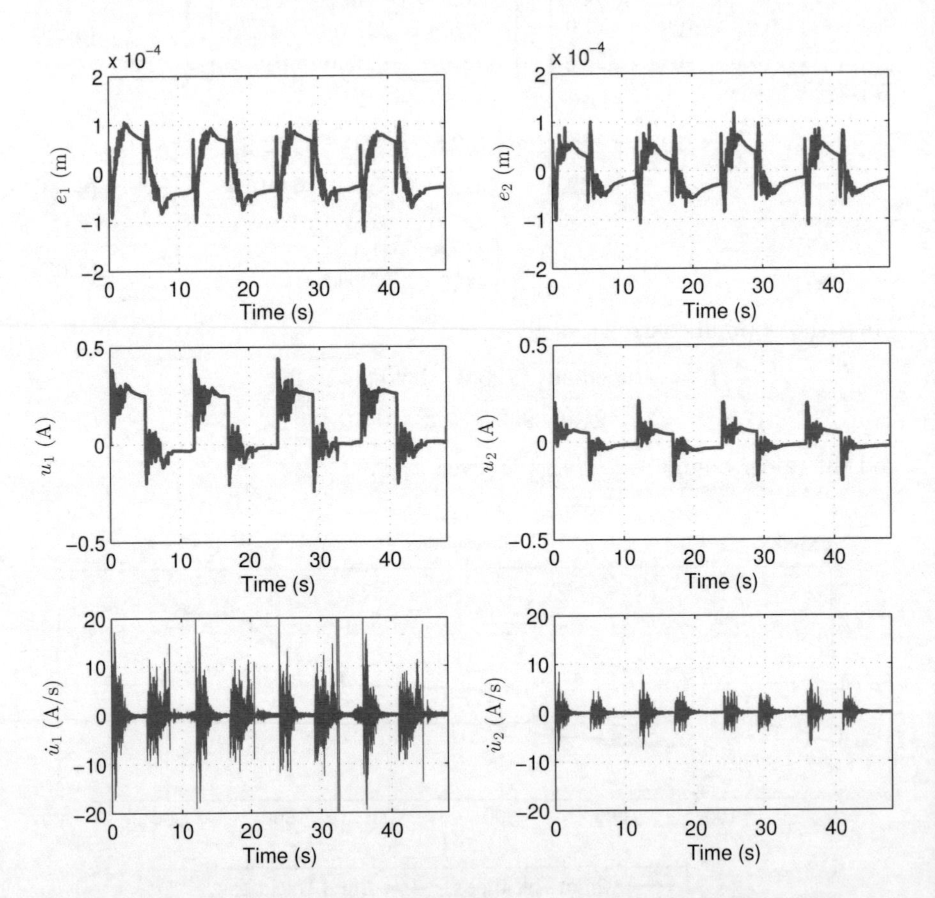

Figure 4.3: Tracking errors, control efforts and their chattering without loading. © [2019] IEEE. Reprinted, with permission, from J. Ma, S.-L. Chen, W. Liang, C. S. Teo, A. Tay, A. Al. Mamun, and K. K. Tan, Robust Decentralized Controller Synthesis in Flexure-Linked H-Gantry by Iterative Linear Programming, IEEE Transactions on Industrial Informatics, vol. 15, no. 3, pp. 1698–1708, 2019.

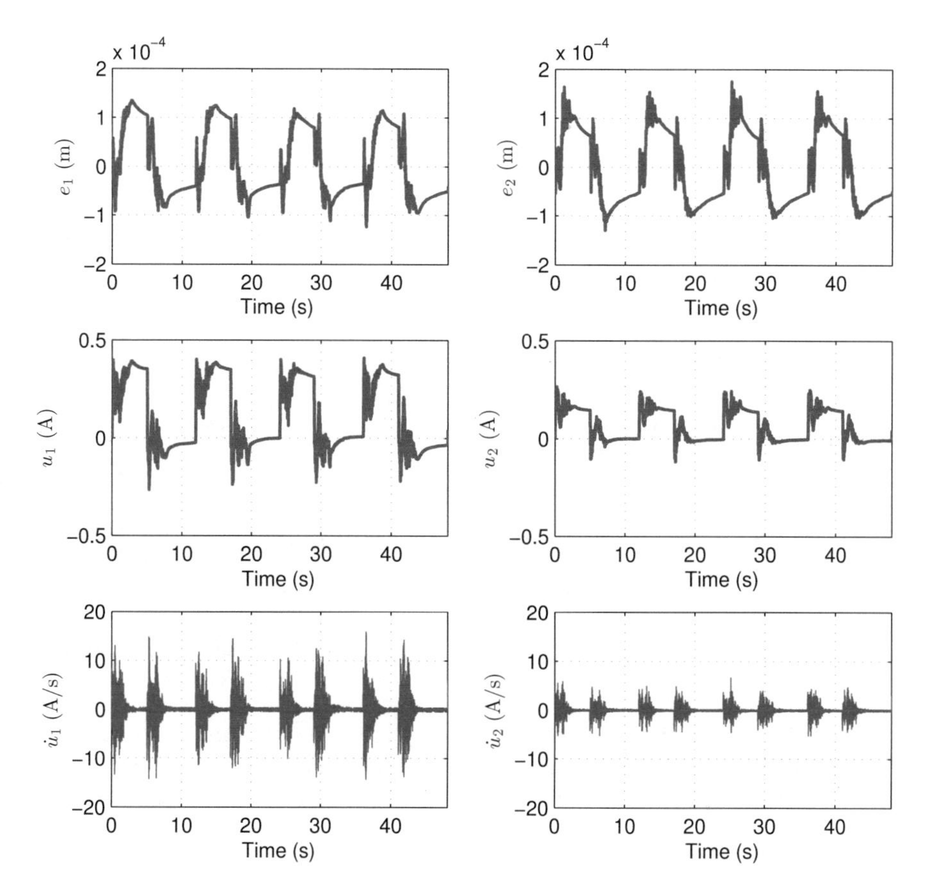

Figure 4.4: Tracking errors, control efforts and their chattering with loading of 0.75 kg on each axis. © [2019] IEEE. Reprinted, with permission, from J. Ma, S.-L. Chen, W. Liang, C. S. Teo, A. Tay, A. Al. Mamun, and K. K. Tan, Robust Decentralized Controller Synthesis in Flexure-Linked H-Gantry by Iterative Linear Programming, IEEE Transactions on Industrial Informatics, vol. 15, no. 3, pp. 1698–1708, 2019.

algorithm, we place a loading of 0.75 kg on each axis of the DHG and retain the same controller gains to conduct the experiment again.

Figure 4.4 demonstrates tracking errors, control efforts as well as their chattering of two axes. In addition, the RMSEs and the MaxAEs in two axes are calculated, and the results are shown in Table 4.2. It can be seen that in the case of no loading applied, the optimal controllers do provide good performance in terms of tracking errors and chattering of control efforts. On the other hand, when the loadings are applied to the system, the tracking performance is slightly worse, but the maximum error is still less than 0.2%

Table 4.2: Comparison of RMSEs and MaxAEs without/with loading © [2019] IEEE. Reprinted, with permission, from J. Ma, S.-L. Chen, W. Liang, C. S. Teo, A. Tay, A. Al. Mamun, and K. K. Tan, Robust Decentralized Controller Synthesis in Flexure-Linked H-Gantry by Iterative Linear Programming, IEEE Transactions on Industrial Informatics, vol. 15, no. 3, pp. 1698–1708, 2019.

Error	Without loading		With loading	
	Axis 1	Axis 2	Axis 1	Axis 2
RMSEs (mm)	0.0543	0.0398	0.0720	0.0758
MaxAEs (mm)	0.1197	0.1185	0.1349	0.1752

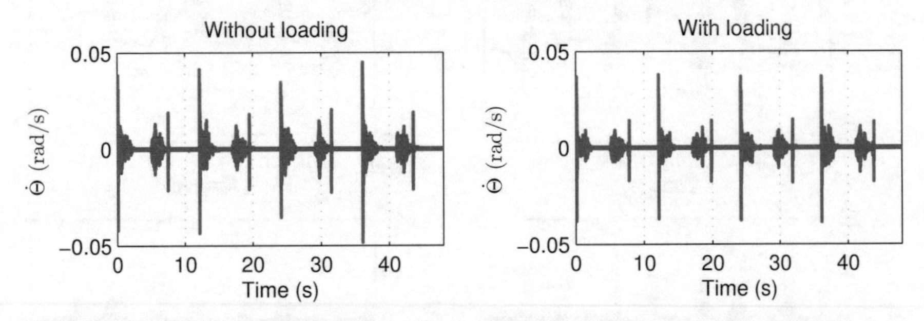

Figure 4.5: Angular velocities measured by laser interferometer. © [2019] IEEE. Reprinted, with permission, from J. Ma, S.-L. Chen, W. Liang, C. S. Teo, A. Tay, A. Al. Mamun, and K. K. Tan, Robust Decentralized Controller Synthesis in Flexure-Linked H-Gantry by Iterative Linear Programming, IEEE Transactions on Industrial Informatics, vol. 15, no. 3, pp. 1698–1708, 2019.

of the overall movement distance, which is within the acceptable range. This indicates the good robust performance of the closed-loop system.

In both cases, the yaw angles between two carriages Θ are measured by the laser interferometer, and the chattering is estimated by differentiation without the use of any low-pass filter. It is clearly shown in Figure 4.5 that the amplitude of the chattering is sufficiently small, and the motion of the horizontal bar is quickly settled without any residual vibration. This validates that the optimization results do achieve precision tracking and minimize the chance of resonant mode excitation.

4.4.4 Summary

In this work, a systematic optimization method is proposed to cater to the real-time computation requirement of the integrated mechatronic design problem in reconfigurable manufacturing systems. The various mechanical and controller constraints, as well as model uncertainties on the state space are converted to the sets on the parameter space. Since system identification cannot be accurate for parallel mechanisms, flexure-linked DHG serves as a special case by using the proposed cutting algorithm, and additional linear constraints on the parameter space are included to preserve the desired decentralized feedback structure in the presence of some uncontrollable but stabilizable states. The optimality and the robustness of such a design approach are successfully validated through comparative experiments on the testbed.

4.5 Conclusion

Since the model uncertainties are not explicitly handled in the linear quadratic based and the $\mathscr{H}_2$-based constrained optimization algorithms, an $\mathscr{H}_2$ guaranteed cost control problem considering convex-bounded model uncertainties is presented, with the objective to minimize the upper bound to the $\mathscr{H}_2$-norm. Since the gain matrix is under certain structural constraints, a series of algorithms is proposed to convert this problem to a decentralized control system design problem without structural constraint. As supported by relevant theoretical results, the constraints from the state space are transformed to the extended parameter space; then all the stabilizing gains that satisfy the structural constraints are parameterized over the intersection of a convex set and a non-convex set defined by a nonlinear real-valued function. Due to the geometrical properties of the above function, numerical procedures can be developed by means of convex programming. Subsequently, a cutting plane based optimization algorithm is developed based on outer-linearization technique, so that a global optimal solution is obtained. The flexure-linked DHG is used as a case study to illustrate the implementation of the proposed algorithm, and the experimental results successfully validate the optimality and the robustness by using the proposed cutting plane based optimization algorithm.

Part II

Data-Based Optimization for Motion Control Systems

5

Reduced-Order Inverse Model Optimization

5.1 Background

Classical model-based control theory requires modeling of systems using either first principles (Teo, Zhu, Chen, Yang, and Pang 2016; Zhao, Liu, He, and Luo 2016; Zhu, Teo, and Pang 2017) or system identification (Chen, Kamaldin, Teo, Liang, Teo, Yang, and Tan 2015; Tan, Liang, Huang, Pham, Chen, Gan, and Lim 2015). To obtain the models for unknown systems can sometimes be difficult and time-consuming. Moreover, inaccurate model-based controller design may not provide satisfactory performance. To solve this issue, data-based methods are developed, which make use of the data generated from the actual system instead of relying upon the plant model (Yin, Li, Gao, and Kaynak 2015; Hou and Wang 2013; Hou and Xu 2009). This direct control-objective-oriented design approach without the intermediate step of system modeling is suitable for many industrial applications (Jiang, Zhu, Yang, Hu, and Yu 2015; Li, Zhu, Yang, and Hu 2015; Li, Zhu, Yang, Hu, and Mu 2017; Radac and Precup 2015).

As a typical data-based control scheme, IFT makes use of a finite set of closed-loop experiment data to construct an unbiased estimation of the gradient of some control performance criteria (Hjalmarsson, Gevers, Gunnarsson, and Lequin 1998; Hjalmarsson 2002; Ren, Xu, and Li 2015). This estimated gradient is then used to tune the parameters of a fixed structure feedback controller iteratively. Convergence to the local minimum of the cost function value is guaranteed under certain conditions, and the detailed proof is given in (Hjalmarsson, Gunnarsson, and Gevers 1994). Various extensions of IFT include its applications to nonlinear systems (Hjalmarsson 1998), frequency domain tuning (Kammer, Bitmead, and Bartlett 2000), fixed structure feedforward tuning (van der Meulen, Tousain, and Bosgra 2008; Stearns, Mishra, and Tomizuka 2008), iterative tuning of internal model controllers (Rupp and Guzzella 2010), and mixed-sensitivity control (Formentin and Karimi 2013), etc.

In reference tracking applications, ILC (Bristow, Tharayil, and Alleyne 2006; Hu, Wang, Zhu, Zhang, and Liu 2016; Chi, Liu, Hou, and Jin 2015) offers high performance, but it comes at the cost of less flexibility in dealing with varying tasks, as ILC has strict requirements for set-point repetition(Heertjes 2016). As a result, the conventional feedforward controller is still widely adopted in the industry(Liu, Tan, Teo, Chen, and Lee 2013; Clayton,

Tien, Leang, Zou, and Devasia 2009). It effectively compensates the inherent lag of the feedback controller in dealing with tracking problems (Liu, Tan, Chen, Teo, and Lee 2013; Zhu, Pang, and Teo 2016), and forms a 2-DOF control (Liu, Tan, Chen, Huang, and Lee 2012). In many low-cost industrial applications such as pick-and-place and tray-indexing, indirect-dive actuators are commonly used. In such applications, low frequency disturbances, such as Coulomb friction (Chen, Tan, and Huang 2012) and mechanical backlash (Li, Chen, Teo, Tan, and Lee 2015), are quite severe, and the 2-DOF control may not offer satisfactory performance.

To deal with these low frequency disturbances effectively, a DOB is included in the inner control loop in addition to the original 2-DOF controller (Schrijver and Dijk 2002; Sariyildiz and Ohnishi 2015b; Yang, Zheng, Li, Wu, and Cheng 2015; Li, Su, Wang, Chen, and Chai 2015; Sariyildiz and Ohnishi 2015a; An, Liu, Wang, and Wu 2016). The DOB can attenuate external disturbances as well as disturbances resulting from the model mismatch, while the outer loop feedback controller can be designed based on the nominal model. This separation property(Schrijver and Dijk 2002) enables an independent design for reference tracking and disturbance rejection, forming a 3-DOF composite control structure. Successful applications of this 3-DOF control structure can be found in (Kempf and Kobayashi 1999; Yan and Shiu 2008). It is worth mentioning that, although the system itself has multiple flexible modes, the order of the nominal inverse model used in both the DOB and the feedforward controller is in general not as high as the order of the system itself. It is determined by the requirement of closed-loop bandwidth as well as the order of the reference trajectory.

In this chapter, we propose a data-based optimization method for the reduced-order inverse model used in both the DOB and the feedforward controller in a 3-DOF control structure. This brings us a 3-DOF control structure with optimally tuned inverse system model, with respect to the actual system instead of the system model. Notice that the proposed method differs from IFT as well as auto-tuning methods because the feedback controller is not tuned but fixed. Furthermore, compared with the ILC, the proposed approach makes no change to the structure of the controller but only tunes its parameters. This feature renders the data-based optimization approach potentially more suitable for fixed structure industrial controllers.

5.2 Overview of the 3-DOF Control Structure

Suppose a typical $(N+1)$-mode motion system is $y = P(s)(u_0 + d)$, and $y_m = y - n$, where

$$P(s) = \underbrace{\frac{1}{m_{\mathrm{t}} s(s+b)}}_{P_{\mathrm{r}}(s)} + \underbrace{\sum_{i=1}^{N} \frac{k_i}{m_{\mathrm{t}}(s^2 + 2\zeta_i \omega_i s + \omega_i^2)}}_{P_{\mathrm{f}}(s)} + \tilde{P}(s), \tag{5.1}$$

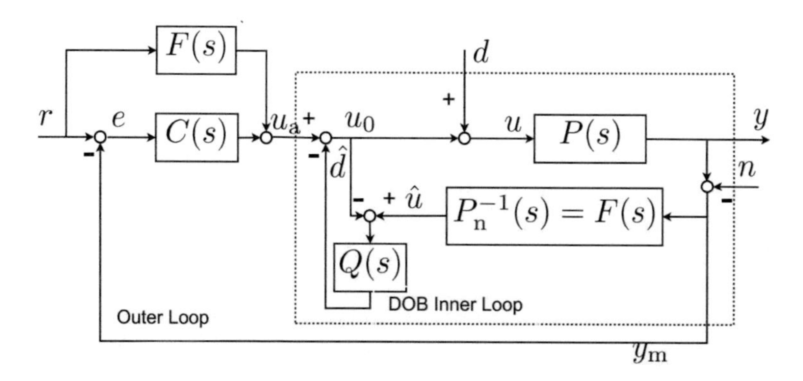

Figure 5.1: 3-DOF control structure. © [2017] IEEE. Reprinted, with permission, from X. Li, S.-L. Chen, C. S. Teo, and K. K. Tan, Data-Based Tuning of Reduced-Order Inverse Model in Both Disturbance Observer and Feedforward with Application to Tray Indexing, IEEE Transactions on Industrial Electronics, vol. 64, no. 7, pp. 5492–5501, 2017.

it is controlled by the 3-DOF control structure as shown in Figure 5.1, where m_{t} is the total moving mass, b is the rigid body damping, k_i, ζ_i and ω_i represent the modal gain, the damping ratio and the natural frequency of i-th mode, and $\tilde{P}(s)$ represents the unmodeled dynamics. This 3-DOF control structure is composed of a feedback controller $C(s)$, a feedforward controller $F(s)$, as well as a DOB included in the inner loop to enhance the disturbance rejection. The key idea of the DOB is to use $P_{\mathrm{n}}^{-1}(s)$ to give an estimation $\hat{u}$ for u, and then feedback the difference between $\hat{u}$ and u_0, which is an estimation of the disturbance d. A low pass filter $Q(s)$ is introduced to reduce high frequency measurement noises and make $Q(s)P_{\mathrm{n}}^{-1}(s)$ realizable. For simplicity, we omit the "(s)" under certain circumstances. As in Figure 5.1, r, e, u_{a}, and y represent the reference, the tracking error, the control input from the outer loop controllers and the system output, respectively. The following frequency responses are realized in the DOB inner loop, when the outer loop is opened(Schrijver and Dijk 2002), where

$$H_{yu_a} = \frac{PP_{\mathrm{n}}}{Q(P - P_{\mathrm{n}}) + P_{\mathrm{n}}}, \tag{5.2}$$

$$H_{yd}^{\mathrm{I}} = \frac{PP_{\mathrm{n}}(1 - Q)}{Q(P - P_{\mathrm{n}}) + P_{\mathrm{n}}}, \tag{5.3}$$

$$H_{yn}^{\mathrm{I}} = \frac{PQ}{Q(P - P_{\mathrm{n}}) + P_{\mathrm{n}}}, \tag{5.4}$$

with the superscript "I" in (5.3) and (5.4) standing for the inner loop. From (5.2)–(5.4), in the low frequency band, the system behaves like the nominal model and the low frequency disturbances are rejected. In the high frequency band, the system behaves like the original plant, and the high frequency noises are rejected.

Due to the unmodeled higher-order dynamics, as well as the non-minimum-phase characteristics due to the loss of sensor-actuator collocation after a certain frequency, the order of the nominal model P_{n} is generally a reduced-order approximation of the P. Meanwhile, the order of feedforward controller is required to be low, usually from 2nd order to 4th order, so that we only need to define the reference trajectory up to 4th order. For instance, implementing a 2nd-order non-causal feedforward controller requires preview information of the velocity as well as the acceleration, and higher-order feedforward controllers require the jerk and even the snap (Lambrechts, Boerlage, and Steinbuch 2005). This low-order feedforward is usually done by either modal truncation or approximation of flexible modes in a lump-sum manner (Boerlage, Tousain, and Steinbuch 2004). In this case, for simplicity, we set $F = P_{\mathrm{n}}^{-1}$ to unify the reduced-order models in both feedforward controller and DOB; then after closing the outer loop, the frequency responses from r to y, r to e, d to y and n to y are modified by $C(s)$ as

$$H_{yr} = \frac{P(C+F)}{1+PC-Q(1-PF)}, \tag{5.5}$$

$$H_{er} = \frac{(1-PF)(1-Q)}{1+PC-Q(1-PF)}, \tag{5.6}$$

$$H_{yd} = \frac{P(1-Q)}{1+PC-Q(1-PF)}, \tag{5.7}$$

$$H_{yn} = \frac{P(C+QF)}{1+PC-Q(1-PF)}. \tag{5.8}$$

H_{er} is also defined as reference sensitivity. For this 3-DOF control system, it is also denoted by S_{R}.

In the next section, we will propose data-based iterative procedures to optimize the parameters of $F(s)$ in 3-DOF control structure as shown in Figure 5.1. Before the optimization process, $C(s)$ and $Q(s)$ are chosen first based on the following propositions.

Proposition 5.1 *In the case that only 2-DOF control* $u_0 = u_{\mathrm{a}} = F(s)r + C(s)e$ *is presented while* $\hat{d} = 0$*, and* $C(s)$ *is designed to stabilize* $P_{\mathrm{r}}(s)$*, then the closed-loop system as in Figure 5.1 is stable if*

$$\sup_{\omega} \left| \frac{C(P_{\mathrm{f}} + \tilde{P})}{1+CP_{\mathrm{r}}} \right| < 1. \tag{5.9}$$

Proof of Proposition 5.1: Proposition 5.1 is directly yielded from the small gain theorem, with treating $(P_{\mathrm{f}} + \tilde{P})$ as an additive model uncertainty of P_{r}.

Proposition 5.2 *Suppose the above 2-DOF control* u_{a} *is able to stabilize* $P(s)$*, and subsequently the DOB* $\hat{d} = F(s)Q(s)y_{\mathrm{m}} - Q(s)u_0$ *is introduced with* *low-pass filter $Q(s)$, to form 3-DOF control $u_0 = u_a - \hat{d}$, then the 3-DOF control closed-loop system is stable if*

$$\sup_{\omega} |QS_r| < 1, \tag{5.10}$$

where $S_r = (1 - PF)/(1 + PC)$ is the reference sensitivity for 2-DOF control closed-loop system.

Proof of Proposition 5.2: The closed-loop stability of the 3-DOF system is equivalent to the stability of $S_R(s) = H_{er}(s)$, and it is easy to show that,

$$S_R(s) = \frac{S_r(1 - Q)}{1 - QS_r}. \tag{5.11}$$

Thus, the stability of $S_R(s)$ is granted by the stability of $S_r(s)$, $(1 - Q)$ and $1/(1 - QS_r)$. Obviously, S_r is stable, since the 2-DOF closed-loop control system is stable. Intuitively, $(1 - Q)$ is stable. And the stability of $1/(1 - QS_r)$ is granted by the small gain theorem, if $\sup_\omega |QS_r| < 1$. This proves Proposition 5.2.

These two propositions provide guidelines for the design of $C(s)$ and $Q(s)$. Firstly, $C(s)$ is designed to satisfy the stability condition according to Proposition 5.1. Secondly, by Proposition 5.2, when an additional DOB loop is introduced, the robust stability is guaranteed by selecting a low-pass $Q(s)$ with sufficiently low cut-off frequency, since $S_r(s)$ is usually high-pass. The next step is to design and perform data-based optimization for $F(s)$ in both the feedforward controller and DOB, and the design for $C(s)$, $Q(s)$ and $F(s)$ forms the complete 3-DOF control structure.

5.3 Reduced-Order Inverse Model Optimization Algorithm

In this section, a fixed structure inverse system model $F(s)$ with linear parameterization is selected first based on the knowledge about the dynamic behavior of a given plant, in the form of

$$F(s, \rho) = \rho^T \bar{F}(s), \tag{5.12}$$

where ρ is a vector of the parameters and $\bar{F}(s)$ is a vector of parameter independent transfer functions. With the structure of $F(s)$ fixed, we proceed to tune ρ towards its optimal value iteratively in order to minimize the tracking error. Thus, the performance criterion is defined as

$$J({}^{\mathbf{i}}\rho) = ({}^{\mathbf{i}}e({}^{\mathbf{i}}\rho))^T \cdot {}^{\mathbf{i}}e({}^{\mathbf{i}}\rho), \tag{5.13}$$

where ${}^{\mathbf{i}}e({}^{\mathbf{i}}\rho)$ is a vector of measured tracking error at every sampling instance of iteration $\mathbf{i}$, with parameter ${}^{\mathbf{i}}\rho$. In this chapter, the bolded left superscript represents the iteration number. We can now formulate the data-based optimization problem as: given the fixed structure $F(s,\rho)$ in (5.12), use solely closed-loop experiment data to determine the parameter vector ρ that minimizes the cost function in $J(\rho)$ (5.13), i.e. to find

$$\rho^\star = \arg\min_{\rho} J(\rho). \tag{5.14}$$

Following (5.13), the gradient of J with respect to ${}^{\mathbf{i}}\rho$ is given by

$$\nabla J({}^{\mathbf{i}}\rho) = 2(\nabla\, {}^{\mathbf{i}}e({}^{\mathbf{i}}\rho))^T \cdot {}^{\mathbf{i}}e({}^{\mathbf{i}}\rho), \tag{5.15}$$

and the Hessian is given by

$$\nabla^2 J({}^{\mathbf{i}}\rho) = 2(\nabla\, {}^{\mathbf{i}}e({}^{\mathbf{i}}\rho))^T \cdot \nabla\, {}^{\mathbf{i}}e({}^{\mathbf{i}}\rho). \tag{5.16}$$

Gauss-Newton algorithm is used, where

$${}^{\mathbf{i+1}}\rho = {}^{\mathbf{i}}\rho - {}^{\mathbf{i}}\gamma(\nabla^2 J({}^{\mathbf{i}}\rho))^{-1}\, \nabla J({}^{\mathbf{i}}\rho), \tag{5.17}$$

where ${}^{\mathbf{i}}\gamma$ is a positive step size at iteration $\mathbf{i}$. In order to execute this optimization algorithm, computation of $\nabla^{\mathbf{i}}e({}^{\mathbf{i}}\rho)$ and ${}^{\mathbf{i}}e({}^{\mathbf{i}}\rho)$ at the current iteration is needed. ${}^{\mathbf{i}}e({}^{\mathbf{i}}\rho)$ can be easily obtained by directly taking the measurement from the closed-loop experiment. We then estimate $\nabla\, {}^{\mathbf{i}}e({}^{\mathbf{i}}\rho)$ solely from the measurement data, similar to the method that is used in IFT. From (5.6), the gradient of the error is given by

$$\nabla^{\mathbf{i}}e({}^{\mathbf{i}}\rho) = \underbrace{\frac{\partial F({}^{\mathbf{i}}\rho)}{\partial \rho} \cdot \frac{QP - P}{\psi({}^{\mathbf{i}}\rho)} \cdot r}_{G_1({}^{\mathbf{i}}\rho)} - \underbrace{\frac{\partial F({}^{\mathbf{i}}\rho)}{\partial \rho} \cdot \frac{\phi({}^{\mathbf{i}}\rho)}{\psi({}^{\mathbf{i}}\rho)\cdot\psi({}^{\mathbf{i}}\rho)} \cdot r}_{G_2({}^{\mathbf{i}}\rho)}. \tag{5.18}$$

$\phi({}^{\mathbf{i}}\rho)$ and $\psi({}^{\mathbf{i}}\rho)$ are defined as follows for simplification of expression, where

$$\phi({}^{\mathbf{i}}\rho) = QP(Q(PF({}^{\mathbf{i}}\rho) - 1) + 1 - PF({}^{\mathbf{i}}\rho)), \tag{5.19}$$

$$\psi({}^{\mathbf{i}}\rho) = Q(PF({}^{\mathbf{i}}\rho) - 1) + 1 + PC. \tag{5.20}$$

By substituting (5.5) into $G_1({}^{\mathbf{i}}\rho)$, it is realized that $G_1({}^{\mathbf{i}}\rho)$ can be generated from the output y, which is obtained by executing a normal finite time task,

$$G_1({}^{\mathbf{i}}\rho) = \frac{\partial F({}^{\mathbf{i}}\rho)}{\partial \rho} \cdot \frac{Q-1}{C + F({}^{\mathbf{i}}\rho)} \cdot y. \tag{5.21}$$

$G_2({}^{\mathbf{i}}\rho)$, on the other hand, is more difficult to obtain, as it requires an additional special experiment. By comparing $G_2({}^{\mathbf{i}}\rho)$ with (5.6), we can get

$$G_2({}^{\mathbf{i}}\rho) = \frac{\partial F({}^{\mathbf{i}}\rho)}{\partial \rho} \cdot \frac{Q}{C + F({}^{\mathbf{i}}\rho)} \cdot H_{yr} \cdot e. \tag{5.22}$$

Thus, $G_2({}^{\mathbf{i}}\rho)$ can be generated by first obtaining the tracking error e during normal operation and then feeding it as a new reference signal to the closed-loop system, which is thus referred to as a special experiment. As shown in (5.22), the measurement of the output of the special experiment $H_{yr} \cdot e$ is needed in order to generate $G_2({}^{\mathbf{i}}\rho)$. Notice that P no longer exists on the right hand side of both (5.21) and (5.22). In addition, $\partial F({}^{\mathbf{i}}\rho)/\partial \rho$, Q, C, $F({}^{\mathbf{i}}\rho)$ and y are all available at iteration $\mathbf{i}$. Thus, the estimation of the gradient can be generated by using the closed-loop experiment data, instead of a plant model.

In summary, the experiments needed for the parameter tuning in a single iteration are listed below.

- Experiment 1: Normal experiment.

$$r^{\mathbf{1}} = r, \tag{5.23}$$

$$y^{\mathbf{1}} = H_{yr} \cdot r + H_{yd} \cdot d^{\mathbf{1}} + H_{yn} \cdot n^{\mathbf{1}}, \tag{5.24}$$

$$e^{\mathbf{1}} = (1 - H_{yr}) \cdot r^{\mathbf{1}} - H_{yd} \cdot d^{\mathbf{1}} - H_{yn} \cdot n^{\mathbf{1}}. \tag{5.25}$$

- Experiment 2: Special experiment.

$$r^{\mathbf{2}} = e^{\mathbf{1}}, \tag{5.26}$$

$$y^{\mathbf{2}} = H_{yr} \cdot e^{\mathbf{1}} + H_{yd} \cdot d^{\mathbf{2}} + H_{yn} \cdot n^{\mathbf{2}}. \tag{5.27}$$

- Experiment 3: Normal experiment.

$$r^{\mathbf{3}} = r, \tag{5.28}$$

$$e^{\mathbf{3}} = (1 - H_{yr}) \cdot r^{\mathbf{3}} - H_{yd} \cdot d^{\mathbf{3}} - H_{yn} \cdot n^{\mathbf{3}}. \tag{5.29}$$

The bolded right superscript refers to the experiment index within a single iteration. In Experiment 1, $y^{\mathbf{1}}$ is measured and used to generate $G_1({}^{\mathbf{i}}\rho)$. Additionally, the measurement of $e^{\mathbf{1}}$ is taken and used as the reference of Experiment 2. In Experiment 2, the measurement of $y^{\mathbf{2}}$ is taken and used to generate $G_2({}^{\mathbf{i}}\rho)$. In Experiment 3, the measurement of $e^{\mathbf{3}}$ is taken and used to calculate the cost function gradient.

The cost function gradient is estimated using the closed-loop experiment data, so external perturbations can cause an error during this estimation. For this stochastic approximation method to work, the estimation has to be unbiased, where

$$E[\text{est}(\nabla J(\rho))] = \nabla J(\rho). \tag{5.30}$$

To prove the unbiasedness, we have the following assumptions.

Assumption 5.1 *Perturbations in different experiments are independent of each other.*

Assumption 5.2 *Perturbations d and n are zero mean, weakly stationary random variables.*

Then, we have the following theorem.

Theorem 5.1 *For the system* (5.1) *under 3-DOF control configuration, as shown in Figure 5.1, under Assumption 5.1 and Assumption 5.2, the estimation of gradient of cost function in* (5.13) *is unbiased with respect to the parameter ρ in the reduced-order model $F(s, \rho)$ in* (5.12).

Proof of Theorem 5.1: Following (5.21) and (5.22), the estimated gradient of the tracking error is given by

$$\begin{aligned} \text{est}(\nabla e(\rho)) &= \frac{\partial F(\rho)}{\partial \rho} \frac{Q-1}{C+F(\rho)} \cdot y^{\mathbf{1}} - \frac{\partial F(\rho)}{\partial \rho} \frac{Q}{C+F(\rho)} \cdot y^{\mathbf{2}} \\ &= \nabla e(\rho) + w, \end{aligned} \tag{5.31}$$

where

$$\begin{aligned} w = &\frac{\partial F(\rho)}{\partial \rho} \cdot \frac{Q-1}{C+F(\rho)} \cdot (H_{yd} \cdot d^{\mathbf{1}} + H_{yn} \cdot n^{\mathbf{1}}) \\ &- \frac{\partial F(\rho)}{\partial \rho} \cdot \frac{Q}{C+F(\rho)} \cdot (H_{yd} \cdot d^{\mathbf{2}} + H_{yn} \cdot n^{\mathbf{2}}). \end{aligned} \tag{5.32}$$

Since w is contaminated by perturbations in Experiment 1 and Experiment 2 and $e^{\mathbf{3}}$ is contaminated only by perturbations in Experiment 3, with Assumption 5.1 and Assumption 5.2, we have

$$E[w^T \cdot e^{\mathbf{3}}(\rho)] = E[w^T] \cdot E[e^{\mathbf{3}}(\rho)], \tag{5.33}$$

$$E[w^T] = 0. \tag{5.34}$$

Expectation of the cost function gradient estimation is then given by:

$$\begin{aligned} E[\text{est}(\nabla J(\rho))] &= 2E[\text{est}(\nabla e^T(\rho))e^{\mathbf{3}}(\rho)] \\ &= 2E[\nabla e^T(\rho)e^{\mathbf{3}}(\rho)] + E[w^T \cdot e^{\mathbf{3}}(\rho)] \\ &= 2E[\nabla e^T(\rho)e^{\mathbf{3}}(\rho)] + E[w^T] \cdot E[e^{\mathbf{3}}(\rho)] \\ &= \nabla J(\rho) + 0 \cdot E[e^{\mathbf{3}}(\rho)] \\ &= \nabla J(\rho). \end{aligned} \tag{5.35}$$

This proves Theorem 5.1.

This is exactly the reason why Experiment 3 is needed. If the tracking error obtained from Experiment 1, instead of Experiment 3, is used in the cost function gradient estimation in (5.15), then the same perturbation from Experiment 1 would exist in both $\text{est}(\nabla e^T(\rho))$ and $e(\rho)$, which leads to a biased cost function gradient estimation.

Eventually, the data-based optimization algorithm for the reduced-order inverse model of the 3-DOF composite control structure is summarized in Algorithm 5.1.

Algorithm 5.1 Data-based Optimization Algorithm for Reduced-order Inverse Model

- Step 1: Set the iteration number $\mathbf{i} = 0$ and select the initial controller parameter ${}^{\mathbf{0}}\rho$.
- Step 2: Execute normal experiment using the desired reference and measure the output $y^{\mathbf{1}}$ and the tracking error $e^{\mathbf{1}}$.
- Step 3: Evaluate the cost function $J({}^{\mathbf{i}}\rho)$. Stop if the cost function value is satisfactory. Otherwise, proceed to Step 4.
- Step 4: Use the output $y^{\mathbf{1}}$ to generate $G_1({}^{\mathbf{i}}\rho)$ according to (5.21).
- Step 5: Feed $e^{\mathbf{1}}$ as the new reference to the closed-loop system and obtain the output $y^{\mathbf{2}}$ from this special experiment. Generate $G_2({}^{\mathbf{i}}\rho)$ according to (5.22) using the obtained output.
- Step 6: Obtain the gradient of the error $\nabla\, {}^{\mathbf{i}}e({}^{\mathbf{i}}\rho)$ according to (5.18).
- Step 7: Conduct another normal experiment and measure the tracking error $e^{\mathbf{3}}$.
- Step 8: Compute $\nabla J({}^{\mathbf{i}}\rho)$ as well as $\nabla^2 J({}^{\mathbf{i}}\rho)$ according to (5.15) and (5.16), where ${}^{\mathbf{i}}e({}^{\mathbf{i}}\rho)$ is $e^{\mathbf{3}}$ from Experiment 3 and $\nabla\, {}^{\mathbf{i}}e({}^{\mathbf{i}}\rho)$ is obtained from Step 6.
- Step 9: Execute the Gauss-Newton algorithm (5.17).
- Step 10: Set the iteration number $\mathbf{i} = \mathbf{i} + 1$ and proceed to Step 2.

5.4 Simulation Analysis

In this section, we apply the proposed data-based method to improve tracking performance of a 6th-order flexible system, where a low-order inverse system model is optimized.

Example 5.1 *Consider a flexible system in the form of* (5.1) *with two resonant modes, impeded by disturbance d and noise n, and the model parameters are obtained from system identification*

$$\begin{aligned} &m_t = 0.0133, b = 8.0, k_1 = 3.325, k_2 = 0.665, \\ &\zeta_1 = 0.08, \zeta_2 = 0.01, \omega_1 = 971.32, \omega_2 = 9458.3. \end{aligned} \tag{5.36}$$

As mentioned, full-order inversion of P will result in a 6th-order feedforward controller, which requires the trajectory to be up to 6th order. To comply with the current industrial practice of low-order motion trajectory planning, a model-based jerk-feedforward controller is developed by taking care of both rigid-body mode and the lump-sum of flexible modes, yielding

$$F(s) = \frac{\rho_1 s^3 + \rho_2 s^2 + \rho_3 s}{s^2 + 2(\zeta_c + \zeta_d)\omega_c s + \omega_c^2}, \tag{5.37}$$

where $\rho_1 = 2m_t\zeta_d\omega_c$, $\rho_2 = m_t\omega_c^2 + 2m_t\zeta_d b\omega_c$, $\rho_3 = \omega_c^2 m_t b$, and $\omega_c = \sqrt{(\sum_{i=1}^{N}(k_i/\omega_i^2))^{-1}}$ is the equivalent resonant frequency, $\zeta_c = b/(2\omega_c)$ is the equivalent damping of resonant modes, and ζ_d is the artificial damping, which is chosen as $(\sqrt{2}/2 - \zeta_c)$ here. Hence, with the model-based design, $F(s)$ is in the form of

$$F(s) = \frac{\rho_1 s^3 + \rho_2 s^2 + \rho_3 s}{s^2 + 752s + 282440}, \tag{5.38}$$

with parameter vector $\rho = [9.9204 \quad 3845.3 \quad 30127]^T$.

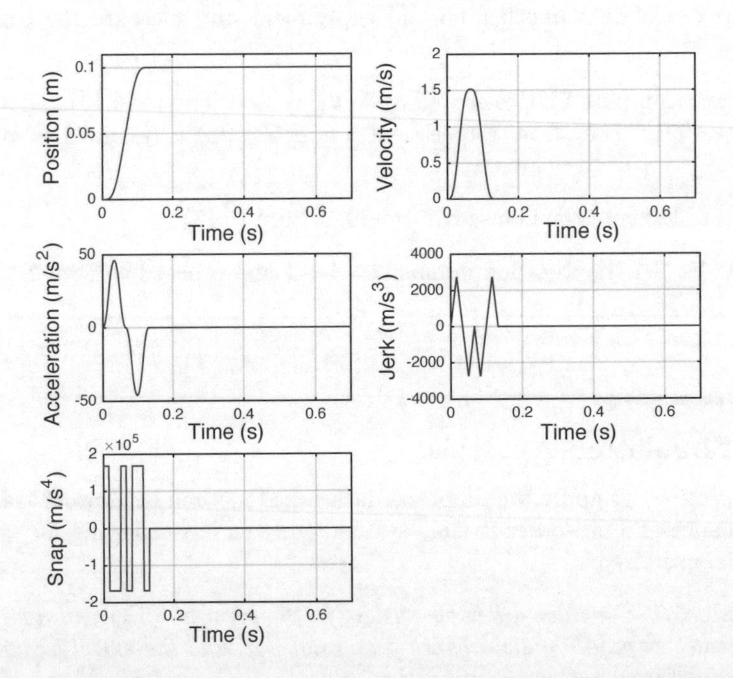

Figure 5.2: Reference profile used in the simulation study. © [2017] IEEE. Reprinted, with permission, from X. Li, S.-L. Chen, C. S. Teo, and K. K. Tan, Data-Based Tuning of Reduced-Order Inverse Model in Both Disturbance Observer and Feedforward with Application to Tray Indexing, IEEE Transactions on Industrial Electronics, vol. 64, no. 7, pp. 5492–5501, 2017.

Figure 5.2 shows the 4th-order set-point used in the simulation. Notice that ω_c is insensitive to modeling errors of high-order resonant modes, so the denominator of (5.38) is fixed here, while we aim to fine-tune the parameter vector ρ in the numerator of (5.38) using our proposed data-based method. In practice, due to inaccurate system identification, ρ is usually not exactly known, so we assume only a rough estimation $\rho = [5 \quad 5000 \quad 20000]^T$ is available. Here, the feedback controller is designed as

$$C(s) = 30 + \frac{8s}{1 + 0.01s}. \tag{5.39}$$

The Q-filter of DOB is designed as a 2nd-order Butterworth filter with cut-off frequency at 100 rad/s to balance the disturbance rejection performance and the high frequency noise reduction

$$Q(s) = \frac{10000}{s^2 + 141s + 10000}. \tag{5.40}$$

The step size ${}^{i}\gamma$ is chosen as 1. By the proposed tuning procedures, the update of parameters is shown in Figure 5.3, as well as the decreasing of the cost function value. This verifies the effectiveness of the proposed data-based optimization method.

Additionally, we find that the optimized parameter vector $\rho = [10.3705 \quad 3836.3 \quad 30529]^T$ is not the same as the vector from model-based design $\rho = [9.9204 \quad 3845.3 \quad 30127]^T$. Notice that by (5.6), if $F(s) = P^{-1}(s)$, $e = 0$. However, the flexible system in this example has a 6th-order model while the feedforward controller is only 3rd order. In this case, the exact match of the inverse dynamics is not possible. Figure 5.4 shows that the controller with the data-based optimization gives lower reference sensitivity in the mid frequency region around 100 rad/s, as compared with the original model-based controller. Here, the reference profile contains mostly low frequency and mid frequency components. In fact, the data-based optimization is able to achieve the optimal trade-off of reference sensitivity between low frequency and mid frequency bands by taking into account the information regarding the reference signal, while the model-based control solely emphasizes on the low frequency bands. The optimized parameters from data-based optimization result in a smaller tracking error as shown in Figure 5.5, and the cost function value is reduced from 1.0578×10^{-8} to 6.3460×10^{-9}.

5.5 Experimental Validation

In this section, the proposed data-based optimization method is applied to the timing belt setup as shown in Figure 5.6, where a digital signal processing card DS1103 is used here for the data acquisition and implementation of

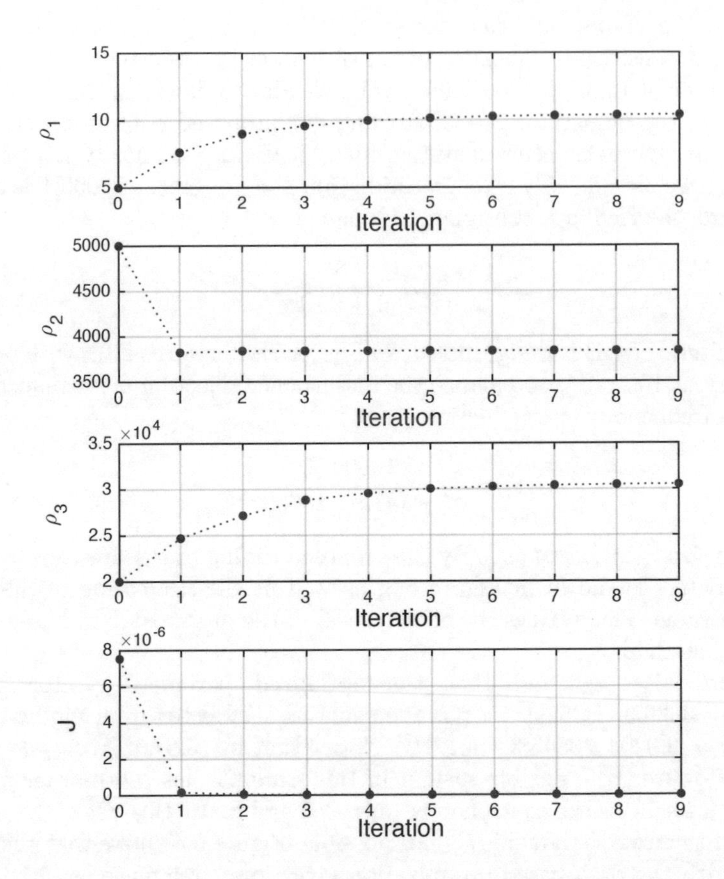

Figure 5.3: Convergence of controller parameters and cost function value reduction in the simulation study.

real-time control. The timing belt is a low-cost indirect-drive solution for efficient conveying and indexing tasks. During belt transmission, the vee-teeth wedge into the groove of the pulley. This is essential for torque transmission capability, and at the same time, it is also a major source of disturbances such as friction and backlash, from which the performance is deteriorated.

5.5.1 Experimental Results

To demonstrate the effectiveness of the proposed data-based method and compare it with the model-based method, system identification is performed at

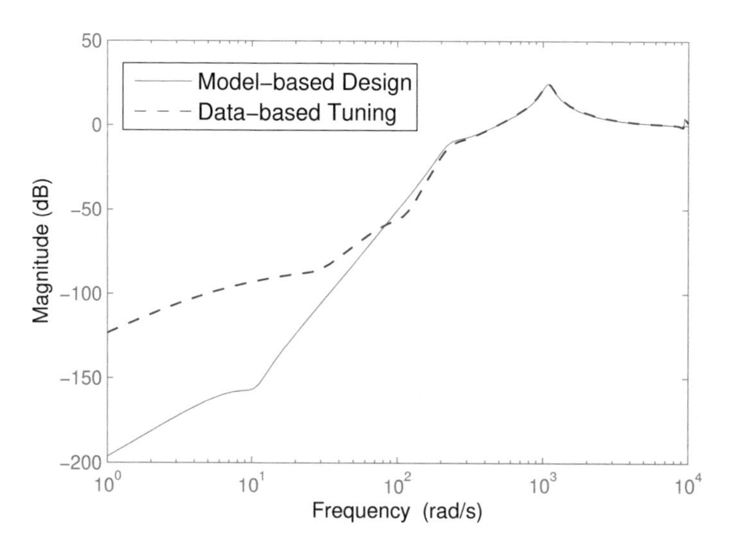

Figure 5.4: Reference sensitivity comparison in the simulation study. © [2017] IEEE. Reprinted, with permission, from X. Li, S.-L. Chen, C. S. Teo, and K. K. Tan, Data-Based Tuning of Reduced-Order Inverse Model in Both Disturbance Observer and Feedforward with Application to Tray Indexing, IEEE Transactions on Industrial Electronics, vol. 64, no. 7, pp. 5492–5501, 2017.

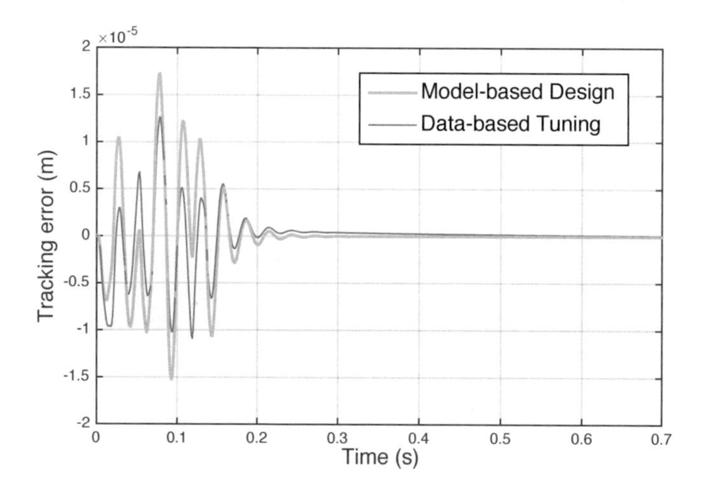

Figure 5.5: Tracking error comparison between model-based and data-based controllers in the simulation study. © [2017] IEEE. Reprinted, with permission, from X. Li, S.-L. Chen, C. S. Teo, and K. K. Tan, Data-Based Tuning of Reduced-Order Inverse Model in Both Disturbance Observer and Feedforward with Application to Tray Indexing, IEEE Transactions on Industrial Electronics, vol. 64, no. 7, pp. 5492–5501, 2017.

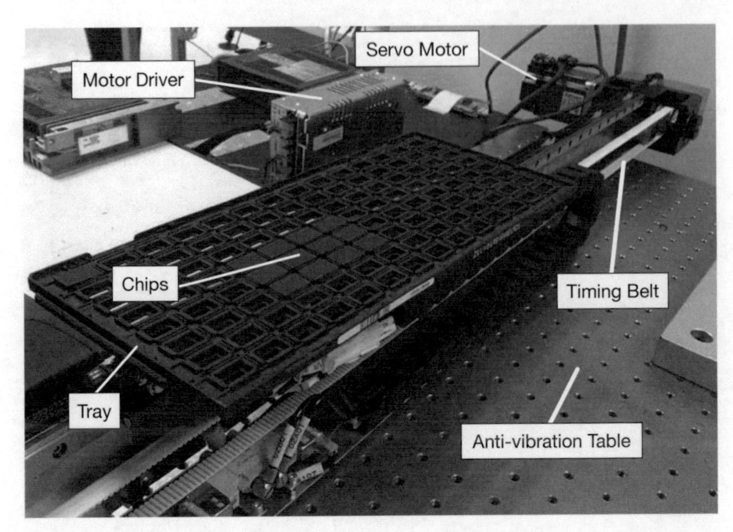

Figure 5.6: Timing belt experimental setup. © [2019] IEEE. Reprinted, with permission, from J. Ma, X. Li, W. Liang, and K. K. Tan, Parameter Space Optimization towards Constrained Controller Design with Application to Tray Indexing, IEEE Transactions on Industrial Electronics, 2019.

the first step by using the driver's utility software, and the result is shown in Figure 5.7. It gives the input dependent Bode plot at the low frequency band, showing a strong influence of disturbances at this band. The system is not operated with high speed and we are mainly interested in the performance below the first anti-resonant frequency. We ignore the phase delay at the high frequency band due to the insufficient sampling rate of the driving utility software, since the sampling rate is sufficiently high at 1 kHz during the subsequent real-time control by DS1103. The model $P(s)$ in the form (5.1) is obtained from Gaussian input with 50% of its maximum torque, and the transfer function parameters are given as (5.36) in Example 5.1.

In the current industrial practice, the nominal model $P_{\mathrm{n}}(s)$ is chosen as the rigid-body mode of $P(s)$, giving the model-based 2nd-order feedforward controller in the form of

$$F(s) = m_{\mathrm{t}}(s^2 + bs) = \rho^T \begin{bmatrix} s^2 & s \end{bmatrix}^T . \tag{5.41}$$

Figure 5.8 shows the reference S-curve profile used in this experiment. The feedback controller is designed as

$$C(s) = 30 + \frac{8s}{1 + 0.01s}, \tag{5.42}$$

and the Q-filter of DOB is

$$Q(s) = \frac{10000}{s^2 + 141s + 10000}. \tag{5.43}$$

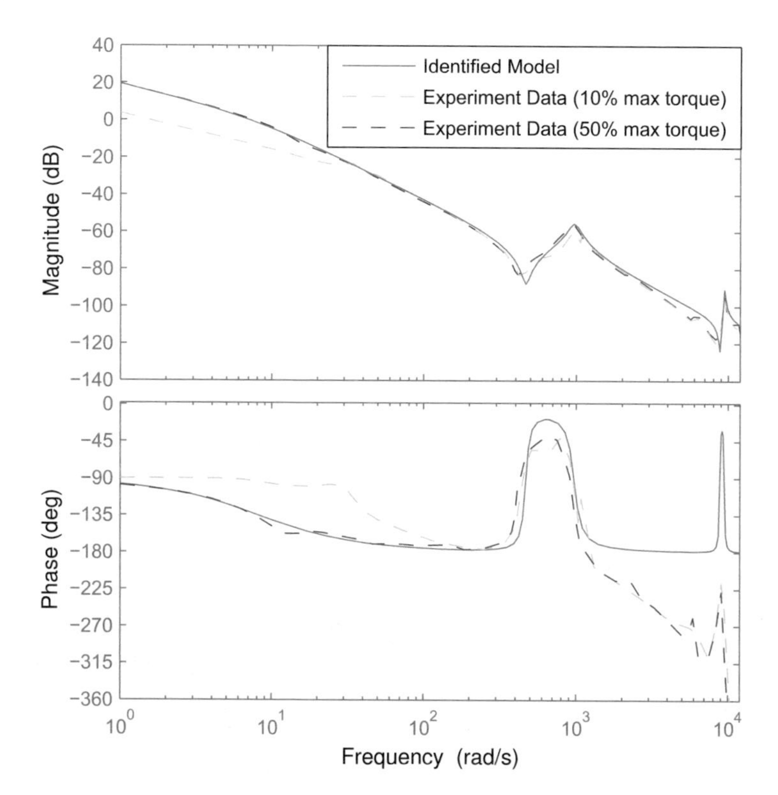

Figure 5.7: Timing belt system identification result. © [2017] IEEE. Reprinted, with permission, from X. Li, S.-L. Chen, C. S. Teo, and K. K. Tan, Data-Based Tuning of Reduced-Order Inverse Model in Both Disturbance Observer and Feedforward with Application to Tray Indexing, IEEE Transactions on Industrial Electronics, vol. 64, no. 7, pp. 5492–5501, 2017.

The step size ${}^{\mathbf{i}}\gamma$ is chosen as 1.

The initial parameter vector is set as $\rho = [0.0133 \quad 0.1064]^T$, according to the system identification result. By using the proposed data-based method, the parameter evolution with respect to the iteration number is plotted in Figure 5.9. Due to the existence of measurement noises in the real-time experiment, to avoid over-training of parameters, the data-based optimization is set to be stopped if the cost function reduction is less than 5% compared to the previous iteration. After 4 iterations, the optimization completes, and the parameter vector ρ converges to the optimal value $\rho = [0.0306 \quad 0.5269]^T$. Also, the reduction of the cost function is plotted in the same figure, and we can see that it is reduced by more than 75% at the same time. Tracking performance improvement is shown in Figure 5.10, where the maximum

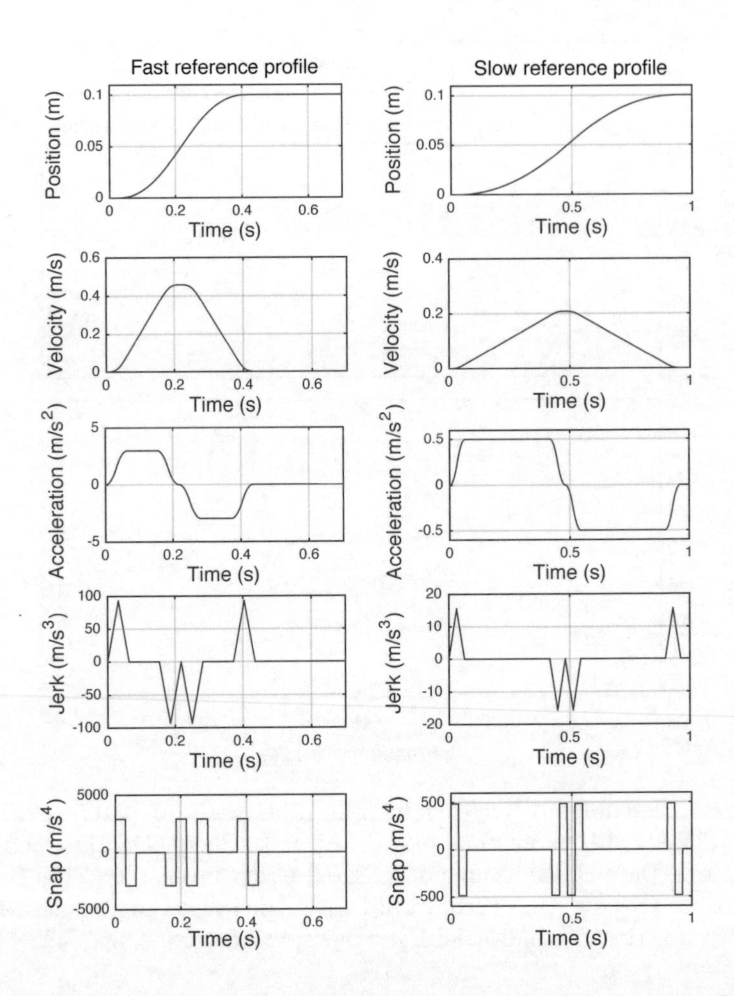

Figure 5.8: Fast and slow reference profiles used in the experiment. © [2017] IEEE. Reprinted, with permission, from X. Li, S.-L. Chen, C. S. Teo, and K. K. Tan, Data-Based Tuning of Reduced-Order Inverse Model in Both Disturbance Observer and Feedforward with Application to Tray Indexing, IEEE Transactions on Industrial Electronics, vol. 64, no. 7, pp. 5492–5501, 2017.

tracking error is improved from over 340 μm to 200 μm, and the tracking error quickly converges to less than 60 μm within 0.6 seconds, showing a significant improvement compared with the model-based counterpart.

5.5.2 Further Analysis and Remarks

In the previous subsection, large input torque with 50% of its maximum value is used to minimize the input torque offset from the friction, resulting in a more

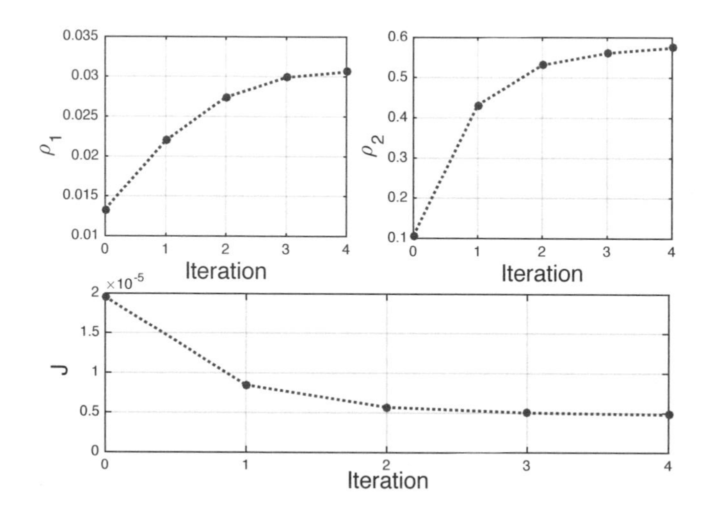

Figure 5.9: Convergence of controller parameters and reduction of cost function value in the experiment for the fast profile. © [2017] IEEE. Reprinted, with permission, from X. Li, S.-L. Chen, C. S. Teo, and K. K. Tan, Data-Based Tuning of Reduced-Order Inverse Model in Both Disturbance Observer and Feedforward with Application to Tray Indexing, IEEE Transactions on Industrial Electronics, vol. 64, no. 7, pp. 5492–5501, 2017.

accurate linear model. In this case, the identified m_t and b should be closed to the actual physical values. However, the value of m_t after optimization is not equal to the initial model-based value. To investigate the reason, we also give the result of system identification using only 10% of the maximum torque input, as shown in Figure 5.7. By comparing these two curves, we observe that in the low frequency band, the identification result shifts downwards when 10% of the maximum torque input is used. Such nonlinearities (due to frictions, backlashes) make it difficult for us to select the optimal parameters used in a linear controller, and the data-based optimization is capable of resolving this issue. If we plot the inverse of $F(s)$ after data-based optimization in Figure 5.11, we can observe that it also shifts downwards. In fact, during the data-based optimization process, the input torque is relatively small due to the small acceleration profile used in the tray-indexing motion. In this case, the inertia "felt" by the actuator is larger than the inertia identified earlier due to the larger effect of friction, yielding a larger equivalent m_t after the optimization. The optimal m_t should be even larger if the reference is slower. This is because in that case, the friction has a comparatively larger effect. This is validated by an additional experiment using a slower reference as shown in Figure 5.8. In this case, the optimal parameter vector converges

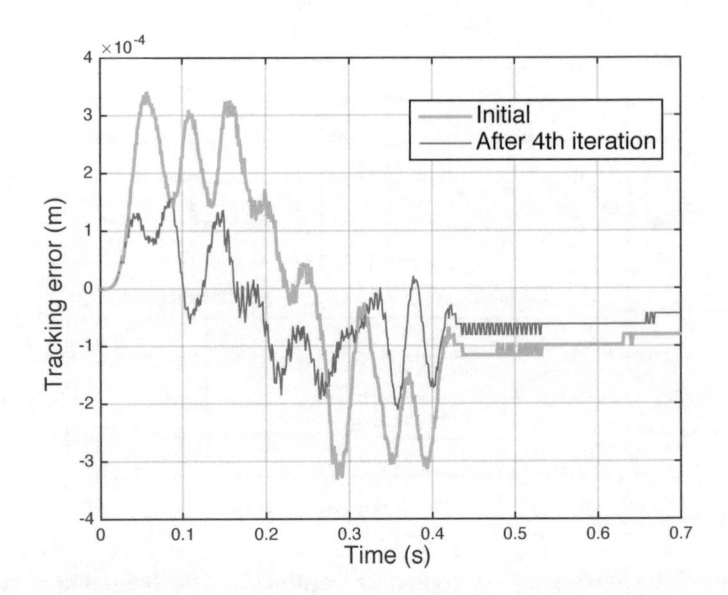

Figure 5.10: Tracking error comparison between model-based (initial) and data-based controllers (after 4th iteration) in the experiment for the fast profile. © [2017] IEEE. Reprinted, with permission, from X. Li, S.-L. Chen, C. S. Teo, and K. K. Tan, Data-Based Tuning of Reduced-Order Inverse Model in Both Disturbance Observer and Feedforward with Application to Tray Indexing, IEEE Transactions on Industrial Electronics, vol. 64, no. 7, pp. 5492–5501, 2017.

to $\rho = [0.0554 \quad 0.7228]^T$, and the inverse of the optimally tuned $F(s)$ is plotted in Figure 5.11, which is shifted further downwards, giving an even larger equivalent m_{t}.

To sum up, since the actual plant is used in the data-based optimization procedures, it can partially deal with some nonlinear effects, such as friction, although the method is based on a linear model. The data-based optimization procedures provide us the best linear controller possible within the parameterized set to match the actual inverse plant dynamics including the nonlinear portion. This feature renders such data-based optimization potentially more attractive for control engineers and researchers.

5.6 Conclusion

In this chapter, we present a data-based optimization method for the reduced-order inverse model in a 3-DOF control structure, which is composed of a

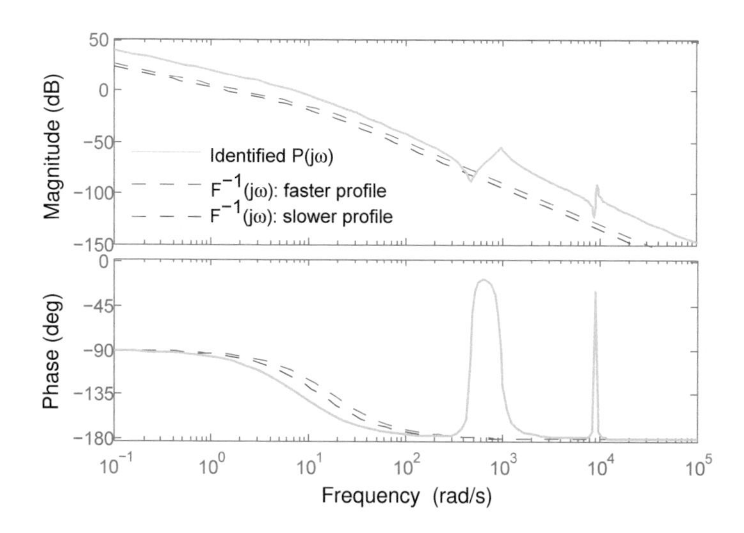

Figure 5.11: Bode diagram of the model and $F^{-1}(s)$ for slow and fast reference profiles. © [2017] IEEE. Reprinted, with permission, from X. Li, S.-L. Chen, C. S. Teo, and K. K. Tan, Data-Based Tuning of Reduced-Order Inverse Model in Both Disturbance Observer and Feedforward with Application to Tray Indexing, IEEE Transactions on Industrial Electronics, vol. 64, no. 7, pp. 5492–5501, 2017.

DOB, a feedforward controller, and a feedback controller. This data-based method supplies an unbiased estimation of the cost function gradient by executing two normal closed-loop experiments and one special closed-loop experiment. This unbiased estimation of the cost function gradient is subsequently used in the Gauss-Newton algorithm to iteratively tune the parameters of a fixed structure controller towards the optimal values. The proposed approach applies to the 2nd-order inverse system model as well as higher-order ones.

The advantages of the proposed data-based optimization algorithm mainly come from three aspects. Firstly, by making use of input and output data only, it avoids the costly process of obtaining an accurate system model. Secondly, it serves as a useful tool for fine-tuning of controllers that have already been determined using the system model. Lastly, the proposed approach achieves improved tracking performance by only tuning the controller parameters without any change to the existing control structure. This is especially useful in the industry because it is quite often that the control engineers are only allowed to modify the controller parameters instead of the controller structure.

6

Reference Profile Alteration and Optimization

6.1 Background

In industrial motion control systems, 2-DOF control is widely used to improve tracking performance. A feedback controller such as the PID controller is used for stabilization of the plant and the feedforward controller is mainly responsible for reference trajectory tracking. The feedback control is inevitably lagging in transient tracking as the tracking error has to occur first before the feedback controller makes the adjustment. Thus, the feedforward control is a useful complement to the feedback control if better tracking performance is required.

In general, the feedforward controller is designed as the inverse of the system dynamics, with a purpose to cancel the system dynamics completely. Ideally, it results in perfect tracking of the reference profile even without the feedback controller, but in reality, it is not possible to get the exact inverse system model as the system is often influenced by disturbances. Thus, the feedback controller is still necessary to further improve the tracking performance and maintain system stability. Practically, disturbances often exist in motion systems and they can be compensated for if measurable. On the other hand, some disturbances are not measurable such as friction and backlashes, etc. As discussed in Chapter 5, a simple effective way to deal with these low frequency disturbances is to use a DOB, which effectively estimates and compensates low frequency external disturbances as well as disturbances resulting from the model mismatch. Both the measured and the estimated disturbances are naturally compensated for by modifying the control input, but in many industrial applications, the control engineers are not allowed to modify the control input directly. The reason is that the commercial controllers are often proprietary and have a closed architecture, so the allowable changes are only to the reference profile and the controller parameters. Given this fact, a new disturbance compensation scheme is proposed which alters the reference profile instead of the control input. However, in this case, the new reference profile is not known beforehand and the non-causality of the feedforward controller becomes a problem. To solve this non-causality problem, a predictive feedforward approach is proposed based on predictions of the reference profile.

The inverse system model used in the feedforward controller and the DOB is generally obtained by system identification, but the performance is very sensitive to modeling inaccuracies. To solve this issue, many data-based approaches are developed to further enhance the performance of the traditional model-based control as described in Chapter 5. This data-based optimization is essentially an add-on feature in nature and it is capable of further enhancing the performance if a higher tracking accuracy is needed, especially when the system identification is less accurate due to nonlinearities and disturbances, etc.

6.2 Problem Formulation

6.2.1 3-DOF Control Structure with DOB

Suppose a typical motion system

$$y = P(z)(u_0 + d), \quad y_m = y - n, \tag{6.1}$$

is controlled by a 2-DOF control scheme $u = u_0$, and

$$u_0 = F(z)r + C(z)e, \quad e = r - y_{\mathrm{m}}, \tag{6.2}$$

where $P(z)$, $C(z)$ and $F(z)$ represent the motion system under control, the feedback controller and the feedforward controller, respectively. For simplicity, the sample-shifting symbol "(z)" in the transfer functions is omitted in most of the following text. The signals r, e, u, d, n and y represent the reference profile, the tracking error, the control efforts, the disturbances, the noises and the system output, respectively. In this case, the transfer function from the reference profile r to the tracking error e is

$$\frac{e}{r} = \frac{1 - PF}{1 + PC}. \tag{6.3}$$

If $F = P^{-1}$, we expect the tracking error to disappear completely, in the absence of d and n. However, in a number of industrial applications such as pick-and-place and tray indexing, indirect-drive actuators are commonly used. In these applications, the low frequency disturbances are very severe and they deteriorate the tracking performance significantly. To further improve the tracking performance, a DOB is introduced in the inner loop, where

$$\hat{d} = Q(u - P_{\mathrm{n}}^{-1} y_{\mathrm{m}}), \tag{6.4}$$

and P_{n} is a reduced-order, nominal model, and Q is a low-pass filter. As shown in Figure 6.1, a 3-DOF composite control is formed, where

$$u = u_0 - \hat{d}. \tag{6.5}$$

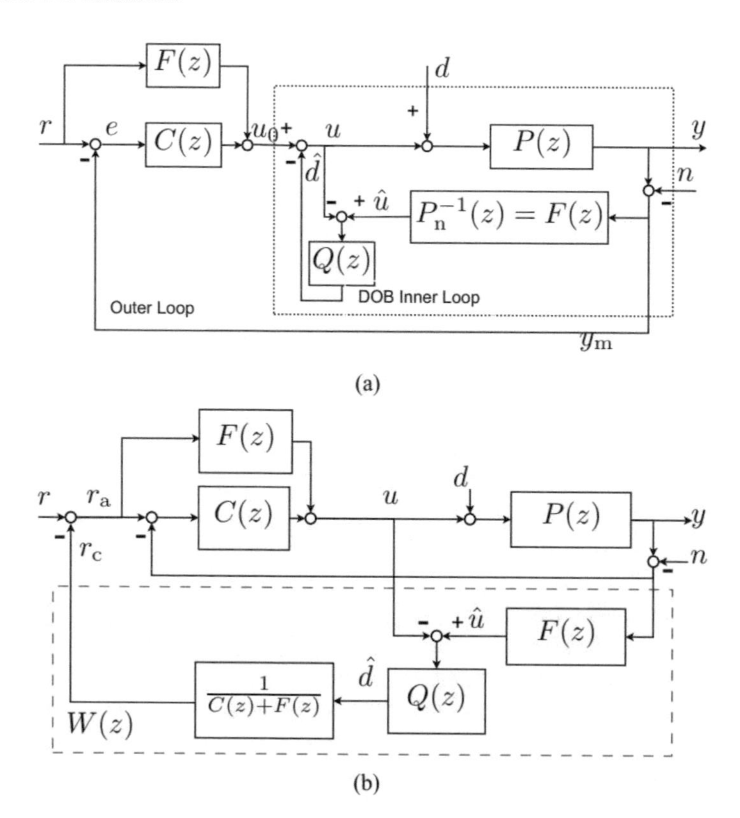

Figure 6.1: Comparison of different disturbance compensation schemes. (a) With DOB. (b) With reference profile alternation. © [2019] IEEE. Reprinted, with permission, from K. K. Tan, X. Li, S.-L. Chen, C. S. Teo, and T. H. Lee, Disturbance Compensation by Reference Profile Alteration with Application to Tray Indexing, IEEE Transactions on Industrial Electronics, vol. 66, no. 12, pp. 9406–9416, 2019.

The key idea of the DOB is to use the inverse of nominal model P_{n}^{-1} to provide an estimation of net control effort $\hat{u}$, and then use it to obtain a disturbance estimation $\hat{d}$. It is worth mentioning that Q should have at least the same order as P_{n} to make the DOB implementable. However, most of the current commercial motion controllers have no built-in DOB loop, and changes of the existing control structure are generally not allowed.

6.2.2 Disturbance Compensation by Reference Profile Alteration

To solve the above issue, we propose to compensate the disturbances by altering the reference profile instead of directly changing the control input, as shown in Figure 6.1. The signals r_{a} and r_{c} represent the altered reference profile and the compensated reference profile, respectively, where

$$r_{\mathrm{c}} = (C+F)^{-1}\hat{d}, \quad r_{\mathrm{a}} = r - r_{\mathrm{c}}, \tag{6.6}$$

and $\hat{d}$ is defined in (6.4) with $F = P_{\mathrm{n}}^{-1}$. In other words, the reference profile r is altered according to the changes of the disturbances. Suppose r_{a} is known prior; the transfer functions in Figure 6.1 are

$$H_{yd}(z) = \frac{P(1-Q)}{1+PC-Q(1-PF)}, \tag{6.7}$$

$$H_{yn}(z) = \frac{P(C+QF)}{1+PC-Q(1-PF)}, \tag{6.8}$$

$$H_{yr}(z) = \frac{P(C+F)}{1+PC-Q(1-PF)}, \tag{6.9}$$

$$H_{er}(z) = \frac{(1-PF)(1-Q)}{1+PC-Q(1-PF)}. \tag{6.10}$$

Especially, in the low frequency range, $Q \approx 1$, and we have $H_{yd} \approx 0$, $H_{er} \approx 0$ and $H_{yr} \approx 1$, which indicates good disturbance rejection and tracking capability. In the high frequency range, $Q \approx 0$, so we have $H_{yn} \approx PC/(1+PC)$, which indicates a similar noise attenuation capability as the pure feedback control case. Thus, this reference alteration scheme is expected to compensate for the disturbances effectively thus improving the tracking performance.

In the situation where the disturbance d is measurable, the DOB is no longer necessary, and we can simply replace $\hat{d}$ with the measured one. However, for either the unmeasurable or measurable disturbances, we do not have the preview information of the altered reference profile r_{a} due to unknown future disturbances, while construction of the feedforward control signal requires prior knowledge of r_{a} (Lambrechts, Boerlage, and Steinbuch 2005). Thus, additional effort to predict the future r_{a} is required, and such prediction will certainly introduce additional prediction error. These issues will be handled appropriately in the next section.

6.3 Predictive Feedforward Scheme with Offsetting Mechanism

Suppose the nominal plant is given by

$$P_{\mathrm{n}}(z) = \frac{\sum_{j=1}^{m} p_j z^j}{\sum_{i=1}^{n} q_i z^i} = \frac{N_{P\mathrm{n}}(z)}{D_{P\mathrm{n}}(z)}, \tag{6.11}$$

with $N_{P\mathrm{n}}$ and $D_{P\mathrm{n}}$ being co-prime. Then,

$$F(z) = P_{\mathrm{n}}^{-1} = \frac{D_{P\mathrm{n}}(z)}{N_{P\mathrm{n}}(z)}, \tag{6.12}$$

and it can be decomposed into its causal part $F_{\mathrm{c}}(z)$ and the non-causal part $F_{\mathrm{n}}(z)$, i.e.

$$F(z) = \sum_{i=1}^{n-m} f_i z^i + F_{\mathrm{c}}(z) = F_{\mathrm{n}}(z) + F_{\mathrm{c}}(z), \tag{6.13}$$

where coefficients f_i can be explicitly determined from (6.12). Here, $F_{\mathrm{c}}(z)$ causes no trouble to the implementation, but $F_{\mathrm{n}}(z)$ requires reference information at least $(n-m)$ samples ahead.

6.3.1 Reference Profile Prediction

In tracking of the reference profile without preview information, the reference profile $r_{\mathrm{a}}(k)$ at current time instant k is available but $r_{\mathrm{a}}(k+1), r_{\mathrm{a}}(k+2), \ldots, r_{\mathrm{a}}(k+m-n)$ are not available. In this subsection, we utilize the polynomial extrapolation approach to estimate the future references based on current and previous information of the reference signal. This approach uses the recent reference signal data to fit into a polynomial function, which is then used to estimate future reference signals.

Suppose the altered reference profile r_{a} can be fit by a qth-order polynomial function

$$r_{\mathrm{a}}(k) = \sum_{i=0}^{q} \alpha_i k^i + \varepsilon, \quad \forall k \geq q+1, \tag{6.14}$$

where ε is the fitting error. Its parameter vector $[\alpha_0, \ldots, \alpha_q]$ is obtained by N available data points with $N \geq q+1$ using the least square method. This polynomial function is then used to predict the future references by

$$\hat{r}_{\mathrm{a}}(k+1) = \sum_{i=0}^{q} \alpha_i (k+1)^i, \quad \forall k \geq q+1. \tag{6.15}$$

Notice that the above approach does not work when $k \leq q$ as there are not enough reference data for us to construct the polynomial function in (6.14). In this case, we use the most recent $r(k)$ as the values of its future prediction

$$\hat{r}_{\mathrm{a}}(k+i) = r_{\mathrm{a}}(k), \quad \forall k \leq q, \quad \forall 1 \leq i \leq n-m. \tag{6.16}$$

Runge's phenomenon(Epperson 1987) states that a higher degree of polynomial does not always improve the extrapolation accuracy, so the linear extrapolation ($q = 1$) is adopted here. Also, the number of data points N is an important design parameter to choose in polynomial extrapolation. Large N value gives less fluctuation but a slower reaction in the prediction, while smaller N has opposite effects.

6.3.2 Offsetting Mechanism for Prediction Error

From the above approach, a prediction error is introduced, where

$$\tilde{r}_{\mathrm{a}}(k) = r_{\mathrm{a}}(k) - \hat{r}_{\mathrm{a}}(k). \tag{6.17}$$

It is filtered by the non-causal $F_{\mathrm{n}}(z)$ and acts as an additional source of disturbance. Hence, we aim to design an additional causal $\Gamma(z)$ to filter $\tilde{r}_{\mathrm{a}}(k)$, as shown in Figure 6.2, so that the resulting output turbulence diminishes as quickly as possible. Subsequently, to maintain the internal 2-DOF controller structure, we shift $\Gamma(z)$ outside.

The order of a polynomial is denoted by $O(\cdot)$. Then, regarding the uniqueness of designing Γ, we have the following theorem.

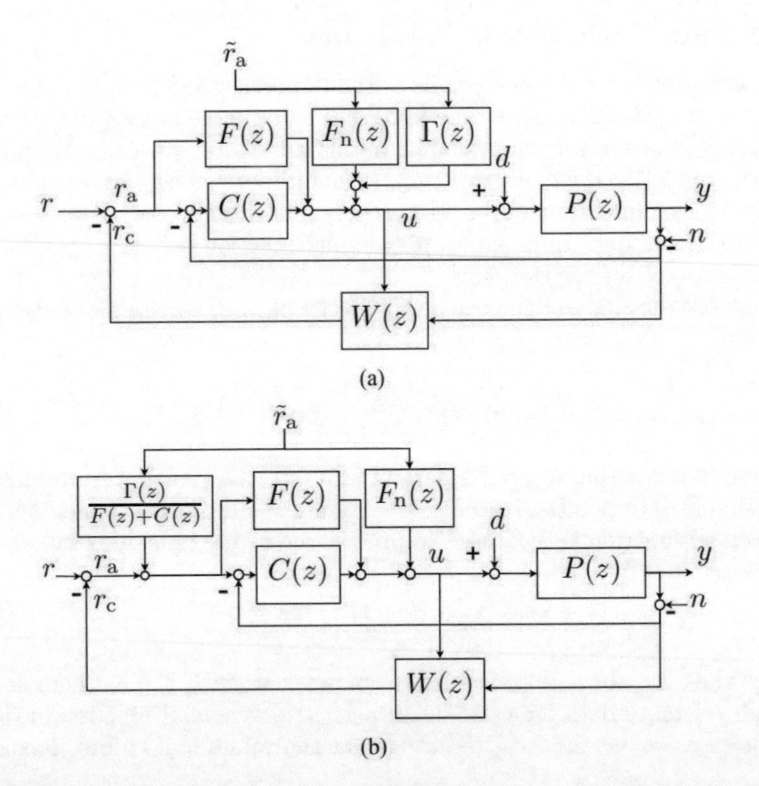

Figure 6.2: Enhanced disturbance compensation caused by $\tilde{r}_{\mathrm{a}}$ by designing $\Gamma(z)$. (a) Before shifting $\Gamma(z)$ to the reference profile. (b) After shifting $\Gamma(z)$ to the reference profile. © [2019] IEEE. Reprinted, with permission, from K. K. Tan, X. Li, S.-L. Chen, C. S. Teo, and T. H. Lee, Disturbance Compensation by Reference Profile Alteration with Application to Tray Indexing, IEEE Transactions on Industrial Electronics, vol. 66, no. 12, pp. 9406–9416, 2019.

Theorem 6.1 *Consider the system $P(z)$ with nominal model $P_{\rm n}(z)$ as defined in (6.11), where $O(N_{P{\rm n}}) = m$, $O(D_{P{\rm n}}) = n$; $F(z)$ is defined in (6.13), where $O(F_{\rm n}) = n - m$. If an additional offsetting mechanism Γ as shown in Figure 6.2 is designed such that $H = P_{\rm n}(\Gamma + F_{\rm n})$ has k deadbeat poles with $k \geq O(N_H) = l \geq 0$, then Γ is uniquely determined by setting $l = k = n - m$.*

Lemma 6.1 *If*

$$P_{\rm n} = \frac{p_1 z + p_0}{z^2 + q_1 z + q_0}, \tag{6.18}$$

then

$$\Gamma = \frac{q_1 + q_0 z^{-1}}{p_1}. \tag{6.19}$$

Define ω_n as the Nyquist sampling frequency, and ω_c as the closed-loop bandwidth by using the 2-DOF control (F and C) structure, and define the tracking error $e = r - y$ in three different configurations, i.e. 3-DOF control scheme as in Figure 6.1, the PAP scheme without and with the offsetting mechanism Γ as in Figure 6.2, to be e_1, e_2 and e_3, respectively. Furthermore, define $\varepsilon_{12} = \|e_2\|_2^2 - \|e_1\|_2^2$, $\varepsilon_{13} = \|e_3\|_2^2 - \|e_1\|_2^2$. Now, regarding the effect of introducing Γ, we have the following theorem.

Theorem 6.2 *Assume $\sup_{\omega \in [0,\omega_n]} |QS_f| \ll 1$, where $S_f = 1 - PF$. $\tilde{r}_a$, r, d, n are zero mean, weak stationary signals in the band $[0, \omega_n]$, and the power spectral density of $\tilde{r}_a$ equals to $\sigma^2_{\tilde{r}_a}$. Define ω_c as the closed-loop bandwidth under $C(z)$, and*

$$\sup_{\omega \in [\omega_c, \omega_n]} \{|PSF_{\rm n}|^2, |PSFH|^2\} \ll \frac{\zeta}{\omega_n - \omega_c}, \tag{6.20}$$

where

$$\zeta = \min\left\{ \int_0^{\omega_c} \left|\frac{F_{\rm n}}{C}\right|^2 d\omega \ , \int_0^{\omega_c} \left|\frac{FH}{C}\right|^2 d\omega \right\}. \tag{6.21}$$

Then, by imposing Γ designed by Theorem 6.1,

1. *$\varepsilon_{13} < \varepsilon_{12}$ if and only if*

$$\int_0^{\omega_c} \left|\frac{FH}{C}\right|^2 d\omega < \int_0^{\omega_c} \left|\frac{F_{\rm n}}{C}\right|^2 d\omega. \tag{6.22}$$

2. *$\varepsilon_{13} \approx 0$ if $P = P_{\rm n}$ and*

$$\frac{\sup_{\omega \in [0,\omega_c]} |H|}{\inf_{\omega \in [0,\omega_c]} |PC|} \ll \frac{2\pi}{\sigma^2_{\tilde{r}_a}}. \tag{6.23}$$

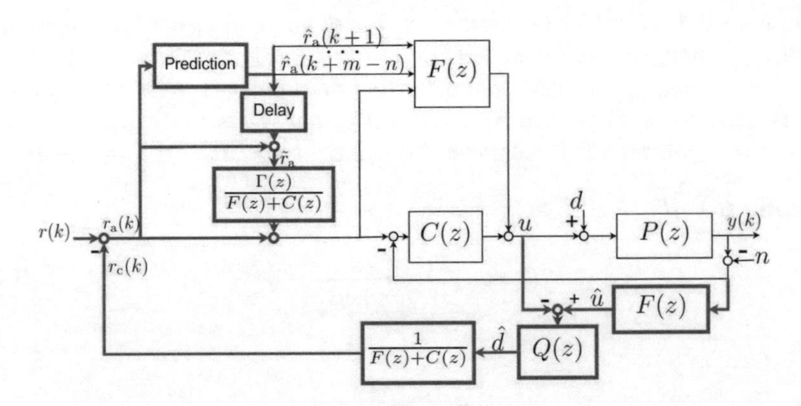

Figure 6.3: Complete implementation diagram with the offsetting mechanism. The dark lines represent the proposed disturbance compensation scheme, whereby no modification is made to the original control structure.

Lemma 6.2 *For $P_n = (p_1 z + p_0)/(z^2 + q_1 z + q_0)$ and $\Gamma = (q_1 + q_0 z^{-1})/p_1$, and where all the assumptions in Theorem 6.2 hold, then $\varepsilon_{13} < \varepsilon_{12}$ if*

$$|e^{j2\omega T_s} + q_1 e^{j\omega T_s} + q_0| > 1, \quad \forall \omega \in [0, \omega_c]. \tag{6.24}$$

The complete implementation diagram is shown in Figure 6.3, where the dark part is the reference profile alteration scheme that is added in addition to the existing control structure. Notice that the original closed-architecture 2-DOF control is not affected and the only change made is the reference profile. Hence, our design objective of implementing a disturbance compensation scheme by reference profile alteration for a closed-architecture controller is achieved.

6.4 Optimization Algorithm for Reference Profile Alteration

From Theorem 6.1 and Theorem 6.2, the introduction of suitable Γ by the profile alteration scheme in Figure 6.3 will make $\|e\|_2^2$ close to the counterpart in the original 3-DOF scheme. In this section, in order to further improve the tracking performance, a data-based optimization approach is implemented to get the optimal parameter vector $\rho = f(p_1, \ldots, p_m, q_1, \ldots, q_n)$ in $F(z, \rho)$, which minimizes the cost function $J(\rho)$, i.e. to find

$$\rho^\star = \arg\min_\rho J(\rho). \tag{6.25}$$

The optimal parameters will then be used in F and Γ as in Figure 6.3. In our problem, we aim to improve the tracking accuracy, so the cost function is defined as

$$J({}^{\mathbf{i}}\rho) = ({}^{\mathbf{i}}e({}^{\mathbf{i}}\rho))^T \cdot {}^{\mathbf{i}}e({}^{\mathbf{i}}\rho), \tag{6.26}$$

and ${}^{\mathbf{i}}e({}^{\mathbf{i}}\rho)$ is the tracking error from iteration $\mathbf{i}$ with parameter ${}^{\mathbf{i}}\rho$. The parameter updating rule is given in (Boyd and Vandenberghe 2004), where

$${}^{\mathbf{i+1}}\rho = {}^{\mathbf{i}}\rho - {}^{\mathbf{i}}\gamma(\nabla^2 J({}^{\mathbf{i}}\rho))^{-1}\, \nabla J({}^{\mathbf{i}}\rho), \tag{6.27}$$

and ${}^{\mathbf{i}}\gamma$ is the step size at iteration $\mathbf{i}$. Furthermore, by (6.26), we have

$$\nabla J({}^{\mathbf{i}}\rho) = 2(\nabla\, {}^{\mathbf{i}}e({}^{\mathbf{i}}\rho))^T \cdot {}^{\mathbf{i}}e({}^{\mathbf{i}}\rho), \tag{6.28}$$

$$\nabla^2 J({}^{\mathbf{i}}\rho) = 2(\nabla\, {}^{\mathbf{i}}e({}^{\mathbf{i}}\rho))^T \cdot \nabla\, {}^{\mathbf{i}}e({}^{\mathbf{i}}\rho). \tag{6.29}$$

With (6.10), the gradient of the tracking error $\nabla^{\mathbf{i}}e({}^{\mathbf{i}}\rho)$ is given by

$$\nabla^{\mathbf{i}}e({}^{\mathbf{i}}\rho) = \underbrace{\frac{\partial F({}^{\mathbf{i}}\rho)}{\partial \rho} \cdot \frac{QP - P}{\psi({}^{\mathbf{i}}\rho)} \cdot r}_{G_1({}^{\mathbf{i}}\rho)} - \underbrace{\frac{\partial F({}^{\mathbf{i}}\rho)}{\partial \rho} \cdot \frac{\phi({}^{\mathbf{i}}\rho)}{\psi({}^{\mathbf{i}}\rho) \cdot \psi({}^{\mathbf{i}}\rho)} \cdot r}_{G_2({}^{\mathbf{i}}\rho)}. \tag{6.30}$$

$\phi({}^{\mathbf{i}}\rho)$ and $\psi({}^{\mathbf{i}}\rho)$ are defined as follows to simplify the expression, where

$$\phi({}^{\mathbf{i}}\rho) = QP(Q(PF({}^{\mathbf{i}}\rho) - 1) + 1 - PF({}^{\mathbf{i}}\rho)), \tag{6.31}$$

$$\psi({}^{\mathbf{i}}\rho) = Q(PF({}^{\mathbf{i}}\rho) - 1) + 1 + PC. \tag{6.32}$$

By substituting (6.9) into $G_1({}^{\mathbf{i}}\rho)$, we can see that $G_1({}^{\mathbf{i}}\rho)$ can be generated from the output y, where

$$G_1({}^{\mathbf{i}}\rho) = \frac{\partial F({}^{\mathbf{i}}\rho)}{\partial \rho} \cdot \frac{Q - 1}{C + F({}^{\mathbf{i}}\rho)} \cdot y. \tag{6.33}$$

Obtaining $G_2({}^{\mathbf{i}}\rho)$ is more difficult, as it requires an additional special experiment. By comparing $G_2({}^{\mathbf{i}}\rho)$ with (6.10), we can get

$$G_2({}^{\mathbf{i}}\rho) = \frac{\partial F({}^{\mathbf{i}}\rho)}{\partial \rho} \cdot \frac{Q}{C + F({}^{\mathbf{i}}\rho)} \cdot H_{yr} \cdot e. \tag{6.34}$$

Thus, $G_2({}^{\mathbf{i}}\rho)$ can be generated by feeding the tracking error e obtained during the first experiment as a new reference signal to the closed-loop system, which is thus referred to as a special experiment. Now $G_1({}^{\mathbf{i}}\rho)$ and $G_2({}^{\mathbf{i}}\rho)$ can be obtained solely based on measurement data; hence a data-based optimization approach is formed.

6.5 Simulation Analysis

In this section, the proposed predictive feedforward method is applied to the same timing belt setup in Chapter 5 and the reference profile is shown in Figure 6.4. In industrial applications, it is a common practice to ignore the parasitic resonant modes and use a 2nd-order system as its nominal model

$$P_{\rm n}(s) = \frac{75}{s^2 + 8s}. \tag{6.35}$$

The nominal model (6.35) is then discretized using the zero-pole matching method with a sampling rate of 1 kHz in the form of (6.18) with $p_1 = p_0 = 3.737 \times 10^{-5}$, $q_1 = -1.992$, and $q_0 = 0.992$. The corresponding feedforward controller is

$$\begin{aligned} F(z) &= \frac{26759(z-1)(z-0.992)}{z+1} \\ &= 26759z + \frac{-80063z + 26545}{z+1}. \end{aligned} \tag{6.36}$$

The causal part of $F(z)$ can be implemented without difficulty, while we need to predict the future value at the next sampling instance in order to implement the non-causal part. As mentioned, this prediction inevitably has error $\tilde{r}_{\rm a}(k)$, and thus $\Gamma(z)$ is designed to partially compensate the error according to (6.19). The feedback controller is designed as

$$C(z) = 30 + 8 \cdot \frac{100}{1 + \frac{0.1}{z-1}}. \tag{6.37}$$

The Q-filter is designed as a 2nd-order Butterworth filter with cut-off frequency at 100 rad/s to balance the disturbance rejection performance and the high frequency noise reduction. Next, we first demonstrate the result for the measurable disturbance case, followed by the unmeasurable disturbance case including the model mismatch case.

6.5.1 Measurable Disturbance Case

Figure 6.5 shows the tracking error comparison when the system is affected by a measurable sinusoidal disturbance with an amplitude of 0.5 V and a frequency of 10π rad/s. We can see the significant performance improvement when the reference profile is altered to compensate for the disturbances. In the same figure, it is shown that the offsetting signal can further reduce the tracking error by partially compensating for the prediction errors. In Figure 6.6, the altered reference profile is plotted against the original reference profile. Also, the prediction error for the altered reference profile is plotted, and we can notice it is quite small in the order of 10^{-6}.

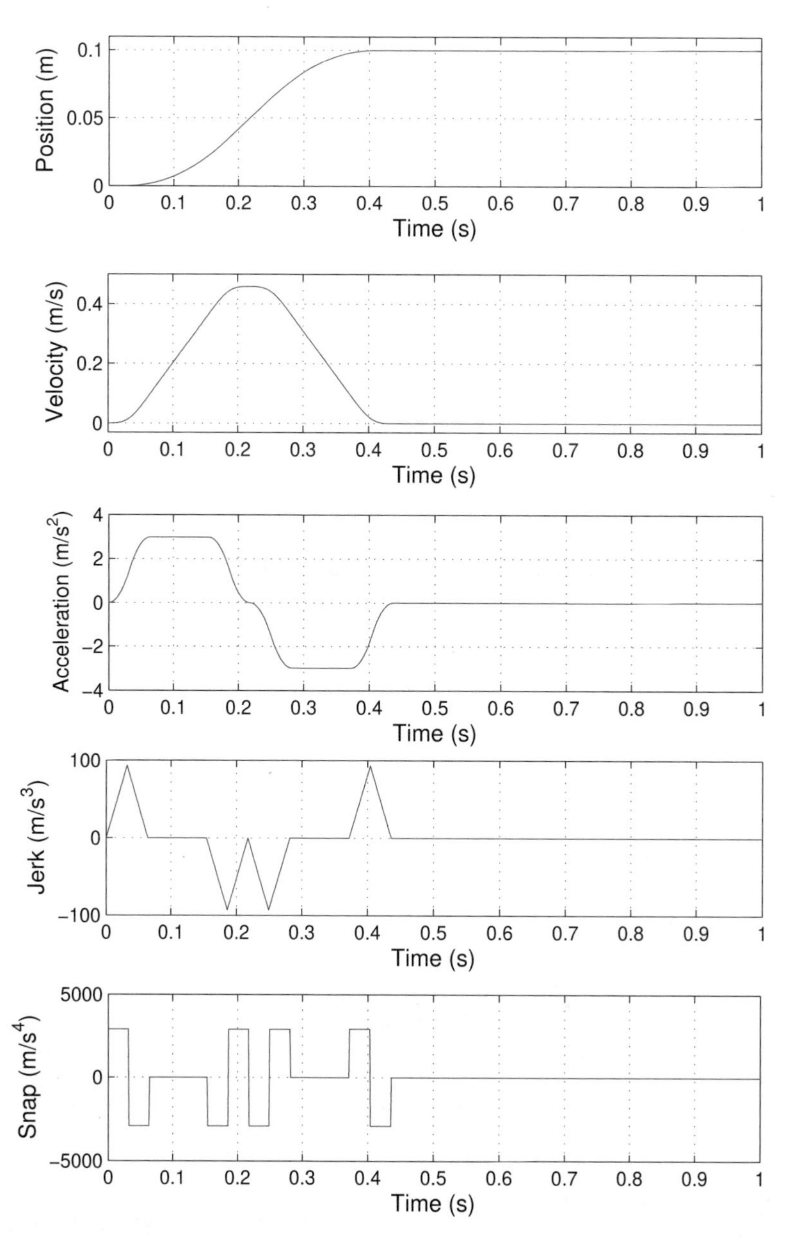

Figure 6.4: 4th-order S-curve used in the simulation and experiment and its velocity, acceleration, jerk, and snap. © [2019] IEEE. Reprinted, with permission, from K. K. Tan, X. Li, S.-L. Chen, C. S. Teo, and T. H. Lee, Disturbance Compensation by Reference Profile Alteration with Application to Tray Indexing, IEEE Transactions on Industrial Electronics, vol. 66, no. 12, pp. 9406–9416, 2019.

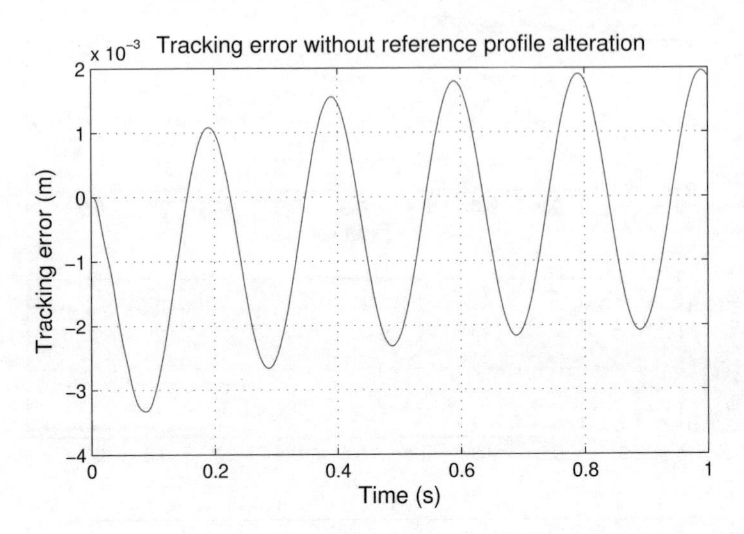

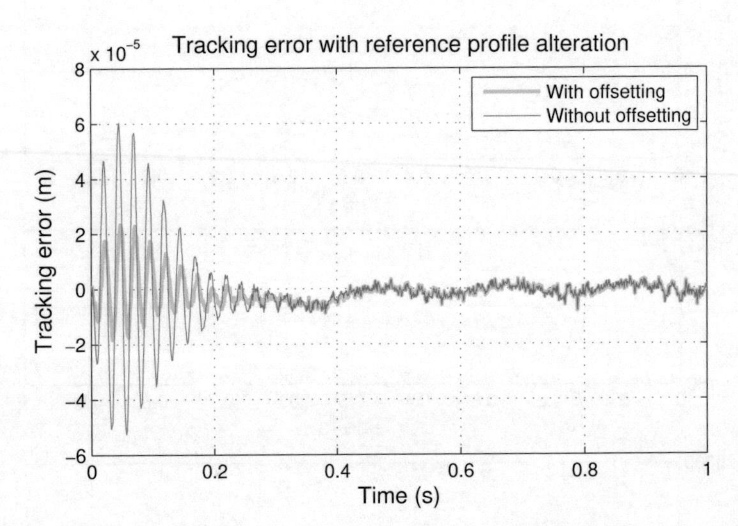

Figure 6.5: Tracking error comparison under a measurable sinusoidal disturbance with an amplitude of 0.5 V and a frequency of 10π rad/s and the effect of the offsetting signal. © [2019] IEEE. Reprinted, with permission, from K. K. Tan, X. Li, S.-L. Chen, C. S. Teo, and T. H. Lee, Disturbance Compensation by Reference Profile Alteration with Application to Tray Indexing, IEEE Transactions on Industrial Electronics, vol. 66, no. 12, pp. 9406–9416, 2019.

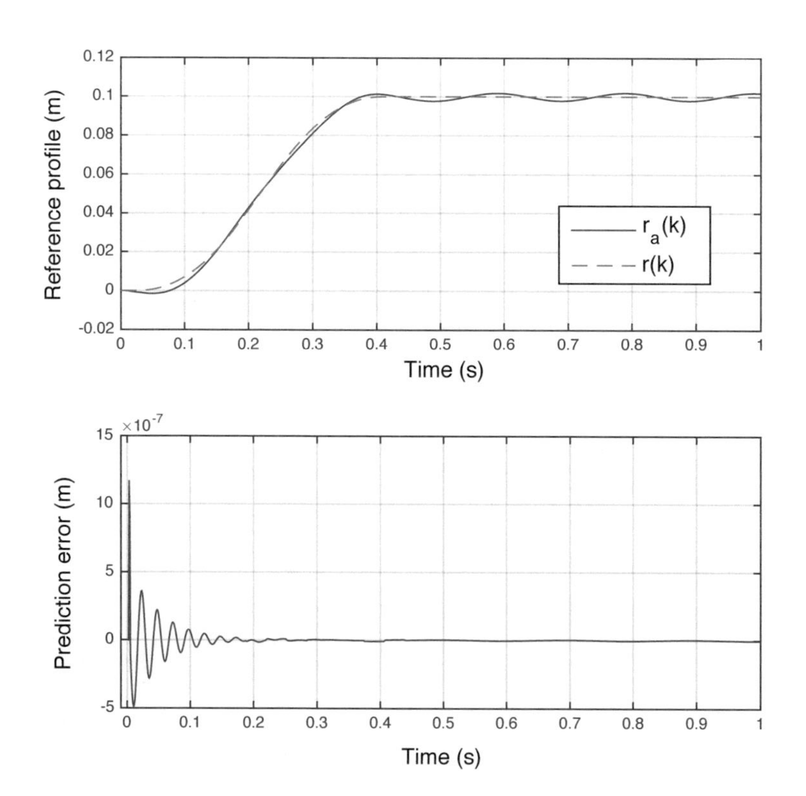

Figure 6.6: Reference profile alteration under measurable sinusoidal disturbances and the prediction error of the altered reference profile. © [2019] IEEE. Reprinted, with permission, from K. K. Tan, X. Li, S.-L. Chen, C. S. Teo, and T. H. Lee, Disturbance Compensation by Reference Profile Alteration with Application to Tray Indexing, IEEE Transactions on Industrial Electronics, vol. 66, no. 12, pp. 9406–9416, 2019.

6.5.2 Unmeasurable Disturbance Case

In the case when the disturbances are not measurable, we make use of a DOB to estimate and compensate for the disturbances using the reference profile alteration scheme as shown in Figure 6.3. Figure 6.7 shows the improvement of tracking accuracy when the system is affected by an unmeasurable step disturbance with an amplitude of 0.1 V. In the same figure, the corresponding reference profile alteration is plotted. Additionally, the reference profile alteration is also able to compensate for the disturbances resulting from model mismatches. In industrial practices, these mismatches could be due to loading variation, machine wear or aging, etc. In the simulation, we assume the plant parameters have the following variation: $m_t = 0.01$, $b = 4.0$.

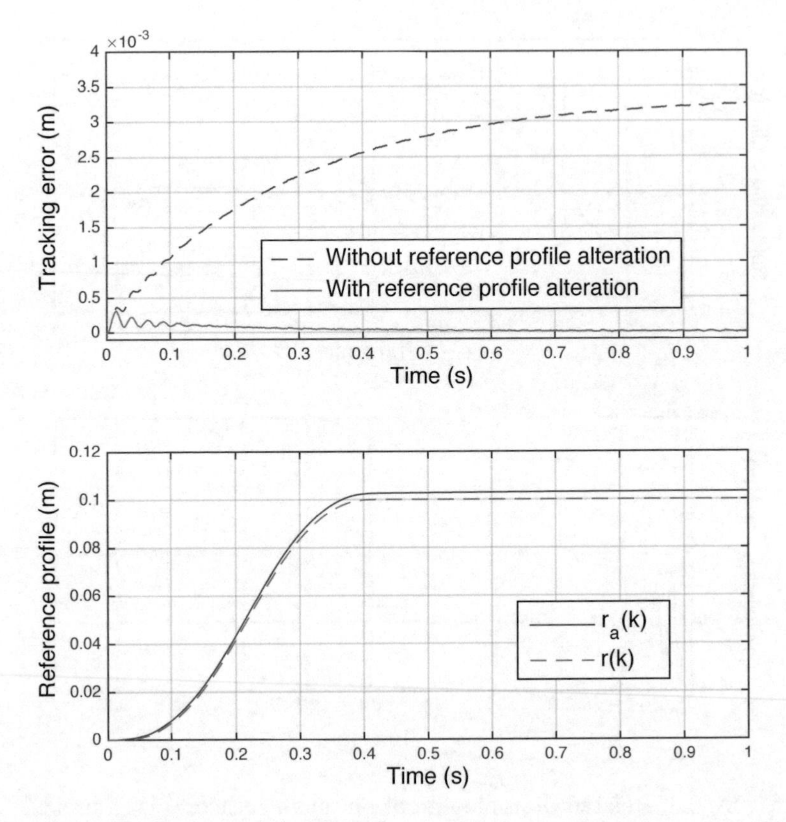

Figure 6.7: Tracking error comparison under a step disturbance with an amplitude of 0.1 V and the corresponding reference profile alteration. © [2019] IEEE. Reprinted, with permission, from K. K. Tan, X. Li, S.-L. Chen, C. S. Teo, and T. H. Lee, Disturbance Compensation by Reference Profile Alteration with Application to Tray Indexing, IEEE Transactions on Industrial Electronics, vol. 66, no. 12, pp. 9406–9416, 2019.

Figure 6.8 shows the improvement of tracking accuracy when such model mismatches exist. Although the reference profile variation is not so obvious, the improvement of tracking accuracy is significant.

6.6 Experimental Validation

In this section, the proposed disturbance compensation scheme by reference profile alteration is applied to the timing belt setup, shown in Figure 5.6. As we already noticed in Figure 5.7, 10% and 50% of maximum input torque give

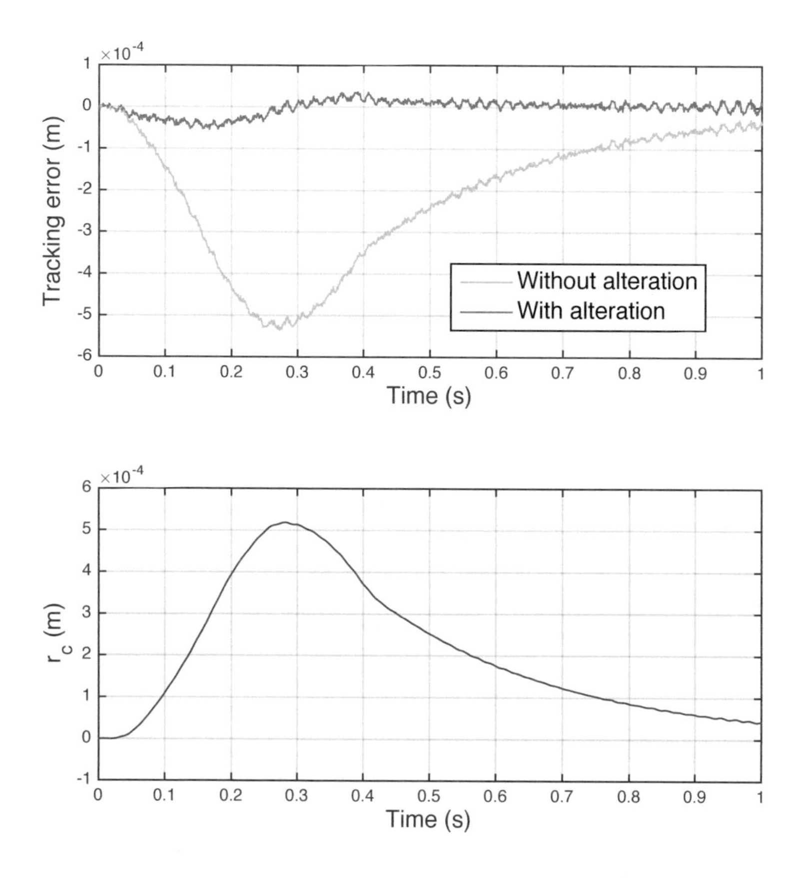

Figure 6.8: Tracking error comparison under model mismatches and the corresponding reference profile alteration $r_c = r - r_a$. © [2019] IEEE. Reprinted, with permission, from K. K. Tan, X. Li, S.-L. Chen, C. S. Teo, and T. H. Lee, Disturbance Compensation by Reference Profile Alteration with Application to Tray Indexing, IEEE Transactions on Industrial Electronics, vol. 66, no. 12, pp. 9406–9416, 2019.

rise to significant differences in terms of the Bode plot in the low frequency band. This inherent nonlinearity is treated as disturbances in our approach and will be compensated by reference profile alteration.

Since the disturbances from friction and backlashes are not measurable, we employ the control structure as in Figure 6.3, which estimates the disturbances and then compensates it through reference profile alteration. The same Q-filter and PD feedback gain are used as in the simulation study. In the case where the traditional 2-DOF controller is implemented without reference profile alteration, we add in an additional integral gain of 200 in the feedback controller to make a fair comparison. Figure 6.9 shows the performance

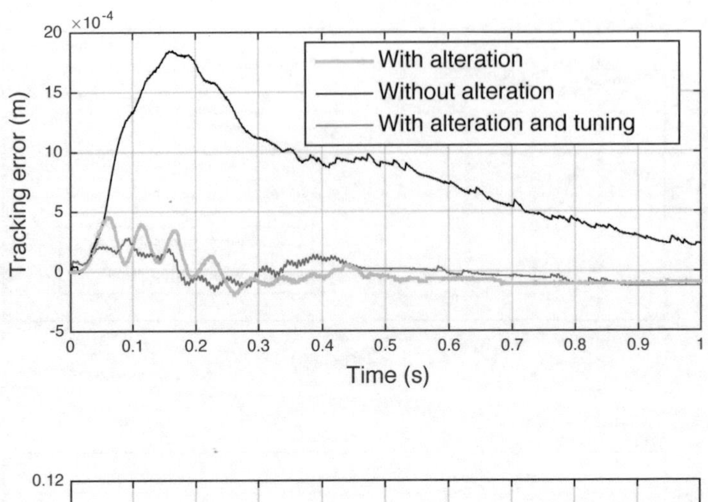

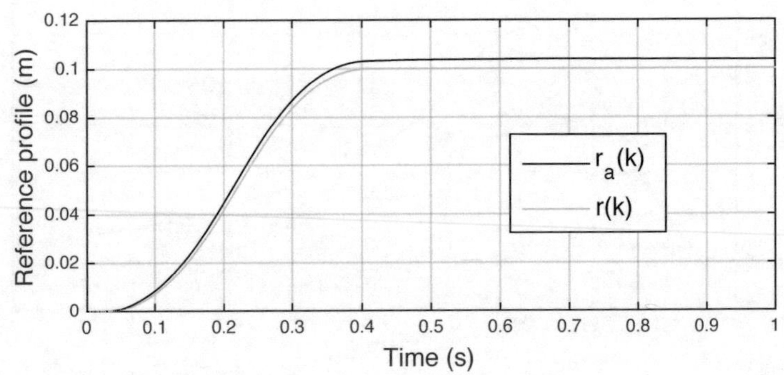

Figure 6.9: Tracking error comparison in real-time experiment and the corresponding reference profile alteration. © [2019] IEEE. Reprinted, with permission, from K. K. Tan, X. Li, S.-L. Chen, C. S. Teo, and T. H. Lee, Disturbance Compensation by Reference Profile Alteration with Application to Tray Indexing, IEEE Transactions on Industrial Electronics, vol. 66, no. 12, pp. 9406–9416, 2019.

improvement when the reference profile alteration scheme is implemented for disturbance compensation. Also, the altered reference profile is plotted against the original reference profile.

Finally, the data-based optimization algorithm in Section 6.4 is implemented. Consider a more generalized form of the inverse of the plant (6.36) as

$$F(z) = \frac{\rho_1(z-1)(z-\rho_2)}{z+1}, \tag{6.38}$$

where ρ_1 and ρ_2 are the two tunable parameters. By executing the data-based optimization procedures, the performance can be further improved as shown

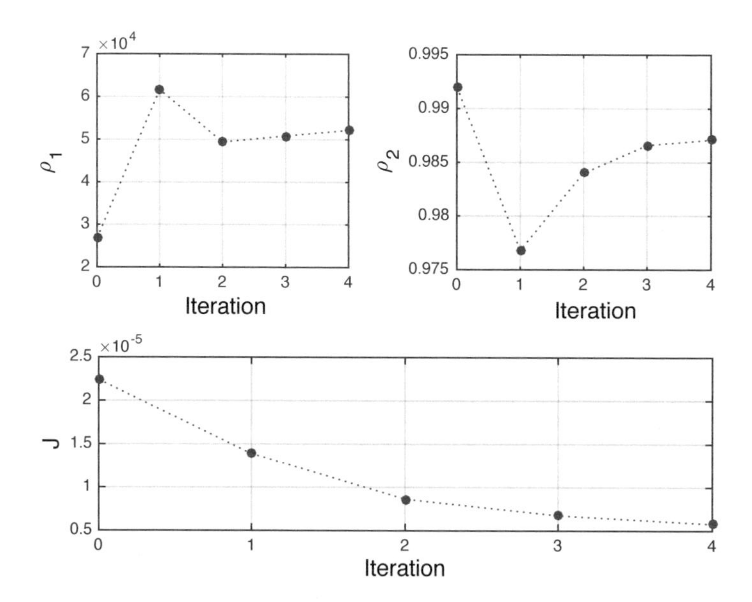

Figure 6.10: Convergence of parameters and reduction of cost function value in the real-time experiment. © [2019] IEEE. Reprinted, with permission, from K. K. Tan, X. Li, S.-L. Chen, C. S. Teo, and T. H. Lee, Disturbance Compensation by Reference Profile Alteration with Application to Tray Indexing, IEEE Transactions on Industrial Electronics, vol. 66, no. 12, pp. 9406–9416, 2019.

in Figure 6.9. The corresponding convergence results for parameter ρ_1, ρ_2 and cost function value J are plotted in Figure 6.10. It can be observed that both ρ_1 and ρ_2 converge to their optimal values and the cost function value is reduced by more than 70%.

6.7 Conclusion

In this chapter, we propose a disturbance compensation scheme using reference profile alteration. Because direct modification to the control input is often not achievable with an existing commercial controller, we aim to compensate for the disturbances by altering the reference profile. Since preview information of the altered reference profile is not available, a predictive feedforward approach is used to deal with the non-causality problem. Additionally, a data-based optimization approach is introduced to further improve the tracking performance. The proposed disturbance compensation scheme is validated in the simulation study, considering measurable disturbances and unmeasurable disturbances including disturbances resulting from the model mismatch. Real-time experimental results based on a timing belt tracking setup further demonstrate the effectiveness of the proposed scheme in industrial applications and showcase the practical appeal of the data-based optimization approach.

7

Disturbance Observer Sensitivity Shaping Optimization

7.1 Background

In motion control systems, disturbances such as friction, backlash, and cogging forces need to be eliminated or at least attenuated, if high accuracy is desired. One simple yet effective way to deal with the low frequency disturbances is to use a DOB, as discussed in the previous two chapters. By estimating the actual disturbances using the inverse system model, the DOB loop can attenuate external disturbances as well as disturbances from the model mismatch. The key benefit of the DOB comes from the fact that it adds disturbance rejection to the feedback control system without affecting the performance of the feedback controller.

The design of DOB is essentially the design of the low pass Q-filter, and the cut-off frequency is carefully selected to balance the low frequency disturbance rejection performance and high frequency noise attenuation as well as robust stability. To further improve the low frequency disturbance rejection capability, a higher cut-off frequency for the Q-filter is necessary. However, the high frequency uncertainties caused by the flexible modes of the motion system as well as other forms of system parameter variations significantly limit the cut-off frequency of the Q-filter. In traditional DOB designs, the Q-filter follows the standard form of Butterworth or binomial design (Kempf and Kobayashi 1999; White, Tomizuka, and Smith 2000; Choi, Yang, Chung, Kim, and Suh 2003). Several advanced designs employ numerical computation methods such as LMIs to design an optimal Q-filter (Wang and Tomizuka 2004; Zhang, Chen, and Li 2008). The optimal Q-filter can also be designed in the framework of a standard $\mathscr{H}_\infty$ control problem (Yun and Su 2013; Yun, Su, Kim, and Kim 2013). Such model-based design is based on the assumption that the system model is reliable and it can represent the actual plant dynamics accurately. In practical applications, such model-based design methods may face challenges when modeling by first principles or identification is difficult or less accurate. With this consideration, the purpose of this work is to develop a data-based approach for sensitivity shaping optimization.

In this chapter, a data-based design of the Q-filter is proposed. By freeing the Q-filter from its traditional forms, 2nd-order or higher-order Q-filters are given additional DOF; thus the shaping of sensitivity functions becomes more flexible. Consequently, the parameters to be determined are no longer simply the cut-off frequency, so a systematic way of selecting them is needed. The design objective is to shape the sensitivity and complementary sensitivity in such a way that the disturbance rejection performance is improved, while robust stability and noise attenuation performance at the high frequency region is not deteriorated. The cost function is designed based on this objective and a data-based approach can be developed to optimize the parameters iteratively.

7.2 Overview of Disturbance Observer-Based Control Systems

The block diagram of a typical DOB inner loop is shown in Figure 7.1, where u, y, d_i and n_i represent the control input signal, the system output, the inherent disturbance, and the inherent measurement noise, respectively. d and n are the injection points of the artificial signal which will only be illustrated in the subsequent section. $P_n^{-1}(s)$ is the inverse of the nominal plant and $Q(s)$ is a low pass filter. The basic idea of the DOB is to use the inverse of the system model to estimate the disturbance to the system and then compensate for it by modifying the control input. The low pass filter $Q(s)$ is introduced to reduce high frequency measurement noise and make $Q(s)P_n^{-1}(s)$ realizable. Hence, the relative degree of $Q(s)$ should be larger than the relative degree of $P_n(s)$. Subsequently, we omit the Laplace variable "(s)" in order to simplify

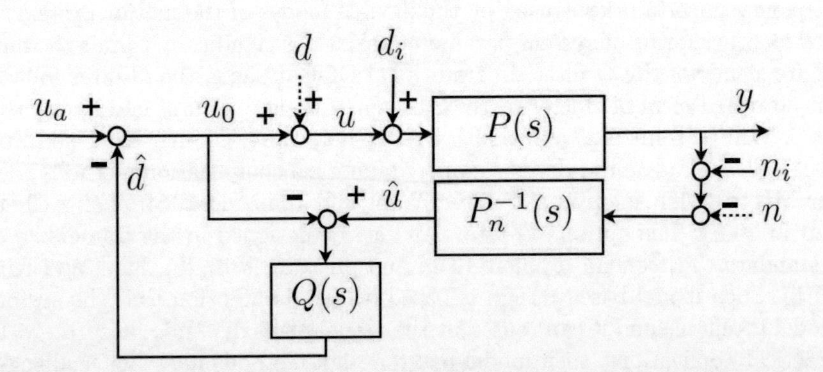

Figure 7.1: DOB inner loop diagram. © [2019] Elsevier. Reprinted, with permission, from X. Li, S.-L. Chen, C. S. Teo, and K. K. Tan, Enhanced Sensitivity Shaping by Data-Based Tuning of Disturbance Observer with Non-Binomial Filter, ISA Transactions, vol. 85, pp. 284–292, 2019.

the equations. The following frequency responses are realized in the DOB inner loop, where

$$H_{yu_a} = \frac{y}{u_a} = \frac{PP_n}{Q(P-P_n)+P_n}, \tag{7.1}$$

$$H^i_{yd} = \frac{y}{d} = \frac{PP_n(1-Q)}{Q(P-P_n)+P_n}, \tag{7.2}$$

$$H^i_{yn} = \frac{y}{n} = \frac{PQ}{Q(P-P_n)+P_n}. \tag{7.3}$$

In the low frequency range, $Q \approx 1$, so we have $H_{yu_a} \approx PP_n/P = P_n$, $H^i_{yd} \approx PP_n(1-Q)/P = 0$, $H^i_{yn} \approx P/P = 1$. Hence, in the low frequency range, the system behaves like the nominal plant and the low frequency disturbances are rejected. In the high frequency range, $Q \approx 0$, and we have $H_{yu_a} \approx PP_n/P_n = P$, $H^i_{yd} \approx PP_n(1-Q)/P_n = P$, $H^i_{yn} \approx PQ/P_n = 0$. Hence, in the high frequency range, the system behaves like the original plant, and the high frequency noises are rejected. Assuming $P = P_n$, the sensitivity function of the inner loop $S_i = 1-Q$ and the complementary sensitivity function $T_i = Q$, so the sensitivity design problem is essentially a Q-filter design problem.

When the feedback controller C is included as shown in Figure 7.2, it is assumed that $P = P_n$; then the sensitivity function is modified as

$$S = \frac{1-Q}{1+PC}, \tag{7.4}$$

and the complementary sensitivity function is modified as

$$T = \frac{PC+Q}{1+PC}. \tag{7.5}$$

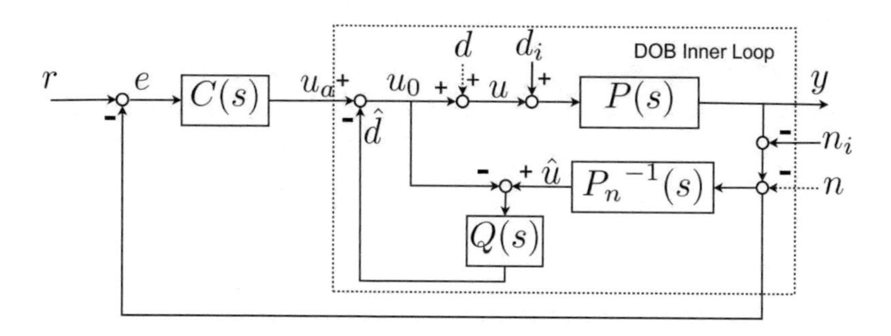

Figure 7.2: 3-DOF control structure. © [2019] Elsevier. Reprinted, with permission, from X. Li, S.-L. Chen, C. S. Teo, and K. K. Tan, Enhanced Sensitivity Shaping by Data-Based Tuning of Disturbance Observer with Non-Binomial Filter, ISA Transactions, vol. 85, pp. 284–292, 2019.

The following frequency responses are realized by the feedback controller and the DOB, where

$$H_{ud} = \frac{u}{d} = \frac{1-Q}{1+PC} = S, \tag{7.6}$$

$$H_{yd} = \frac{y}{d} = \frac{P(1-Q)}{1+PC} = P \cdot S, \tag{7.7}$$

$$H_{yn} = \frac{y}{n} = \frac{PC+Q}{1+PC} = T. \tag{7.8}$$

In the design of the DOB, the disturbance rejection performance depends on the sensitivity function (7.4), while robust stability and noise rejection capability depend on the complementary sensitivity function (7.5). Since $S + T = 1$, reducing both of them at all frequencies is not possible. Considering the fact that disturbances are generally more severe in the low frequency region and noises inherently have higher frequency, the usual approach is to reduce the sensitivity function at the low frequency region and reduce the complementary sensitivity function at the high frequency region.

Moreover, suppose the model uncertainty can be represented in the form of $P(s) = P_n(1 + \Delta(s))$; the robust stability condition for DOB is (Yun and Su 2013):

$$\bar{\sigma}(\Delta(j\omega) \cdot T(j\omega)) < 1, \quad \forall \omega. \tag{7.9}$$

This is a sufficient condition for DOB robust stability as it is derived from the small gain theorem (Doyle, Francis, and Tannenbaum 1992). There are usually more uncertainties in the high frequency region, so a smaller complementary sensitivity function at the high frequency region guarantees better robust stability. In traditional DOB design, the bandwidth is the only parameter that is determined to strike the balance between disturbance rejection and noise attenuation as well as robust stability. The Q-filter commonly follows the standard form of Butterworth or binomial design. Differently, in this work, the Q-filter is designed flexibly without following the standard form, and its parameters are iteratively optimized based on the measurement data. It can be shown that the optimized Q-filter can improve the low frequency disturbance rejection performance while not compromising the high frequency noise attenuation performance and the robust stability condition.

7.3 Sensitivity Shaping Optimization Procedures

A Nth-order Q-filter can be written in the form:

$$Q(s) = \frac{\rho_N}{s^N + \rho_1 s^{N-1} + \rho_2 s^{N-2} + ... + \rho_N}, \tag{7.10}$$

where $\rho_1, \dots, \rho_N$ are the parameters to be determined. In this section, we propose a data-based iterative optimization approach to get the optimal parameters that are able to shape the sensitivity function and complementary sensitivity function in order to achieve the control objectives.

The first task is to design an appropriate cost function which reflects exactly the design purpose. Intuitively, the shaping of sensitivity function S would be the main objective as it is directly related to the disturbance rejection performance. However, the shaping of the complementary sensitivity function T is equally important as it is related to the robust stability. Thus, two experiments need to be conducted separately to feed suitable excitation signals from both the disturbance side point d and noise side point n as shown in Figure 7.2. In Experiment 1, white noise d_1 is injected from the disturbance side and the output y_1 is measured, which is then filtered by a low pass filter. Notice that we have the freedom to choose the cut-off frequency of the low pass filter according to the spectrum of the disturbances. In Experiment 2, white noise n_2 is injected as an excitation signal from the noise side and the output y_2 is measured. According to (7.7) and (7.8), we have $y_1 = P \cdot S \cdot d_1$ and $y_2 = T \cdot n_2$. Now we can define the signal-based cost function as

$$J(\rho) = w_1(L_L \cdot y_1(\rho))^2 + w_2(L_H \cdot y_2(\rho) - L_H \cdot y_0)^2. \tag{7.11}$$

w_1 and w_2 are the weightings that the designer can choose. The low pass filter L_L and high pass filter L_H are included because we are interested only in the low frequency region of S and high frequency region of T. y_0 is the initial output response of Experiment 2, and it is introduced to keep the high frequency portion of T unchanged so that robust stability is not violated. Since y_0 is only measured once at the beginning of the sensitivity shaping process, it is subsequently referred to as Experiment 0. The data-based iterative optimization approach presented next will determine the parameter vector $\rho = [\rho_1, \dots, \rho_N]^T$ that minimizes the cost function, i.e. to find

$$\rho^\star = \arg\min_\rho J(\rho). \tag{7.12}$$

7.3.1 Sensitivity Shaping Algorithm

Similar to the previous chapters, the optimization algorithm based on Newton's method is given by

$$^{i+1}\rho = {}^i\rho - {}^i\gamma(\nabla^2 J({}^i\rho))^{-1}\nabla J({}^i\rho), \tag{7.13}$$

where ${}^i\rho$ and ${}^i\gamma$ are the parameter vector and the positive step size at iteration i, respectively. In order to execute this optimization algorithm, the gradient $\nabla J({}^i\rho)$ and the Hessian $\nabla^2 J({}^i\rho)$ at the current iteration i are needed.

Following (7.11), the gradient of the cost function with respect to the controller parameters can be derived as

$$\nabla J({}^i\rho) = 2w_1 L_L \cdot L_L \nabla y_1^T({}^i\rho) y_1({}^i\rho) + 2w_2 L_H \cdot L_H \nabla y_2^T({}^i\rho)(y_2({}^i\rho) - y_0), \tag{7.14}$$

whereas the approximation of the Hessian of the cost function with respect to the controller parameters is given by

$$\nabla^2 J(^i\rho) = 2w_1 L_L \cdot L_L \nabla {y_1}^T(^i\rho)\nabla y_1(^i\rho) + 2w_2 L_H \cdot L_H \nabla {y_2}^T(^i\rho)\nabla y_2(^i\rho). \tag{7.15}$$

$\nabla y_1(^i\rho)$ and $\nabla y_2(^i\rho)$ are the only unknowns while the rest can be easily obtained from measurement. Inspired by the IFT (Hjalmarsson 2002) , $\nabla y_1(^i\rho)$ and $\nabla y_2(^i\rho)$ can be estimated from measurement data.

With

$$y_1(^i\rho) = \frac{(1-Q(^i\rho))P}{1+PC} \cdot d_1, \tag{7.16}$$

we have

$$\begin{aligned} \nabla y_1(^i\rho) &= -\frac{P}{1+PC} \cdot \frac{\partial Q(^i\rho)}{\partial \rho} \cdot d_1 \\ &= -\frac{1}{1-Q(^i\rho)} \cdot \frac{\partial Q(^i\rho)}{\partial \rho} \cdot y_1(^i\rho). \end{aligned} \tag{7.17}$$

Notice that now ∇y_1 can be directly obtained from measurement data.

Similarly, with

$$y_2(^i\rho) = \frac{PC + Q(^i\rho)}{1+PC} \cdot n_2, \tag{7.18}$$

the gradient of y_2 can be derived as

$$\nabla y_2(^i\rho) = -\frac{1}{1+PC} \cdot \frac{\partial Q(^i\rho)}{\partial \rho} \cdot n_2. \tag{7.19}$$

To obtain ∇y_2, an additional experiment is required, which is denoted by Experiment 3. Notably, n_2 is fed from the d side and u is measured with

$$u_3(^i\rho) = \frac{1-Q(^i\rho)}{1+PC} \cdot n_2. \tag{7.20}$$

Here, u_3 represents u measured in Experiment 3. Using this relationship, we have

$$\begin{aligned} \nabla y_2(^i\rho) &= -\frac{1}{1+PC} \cdot \frac{\partial Q(^i\rho)}{\partial \rho} \cdot n_2 \\ &= \frac{1}{1-Q(^i\rho)} \cdot \frac{\partial Q(^i\rho)}{\partial \rho} \cdot u_3(^i\rho). \end{aligned} \tag{7.21}$$

In this algorithm, the parameters of the Q-filter are optimized to reduce sensitivity in the low frequency region while keeping the high frequency region of complementary sensitivity function relatively unchanged. The mid frequency region is not considered as we assume most disturbances exist in the low frequency region. According to the "waterbed" effect, the disturbance rejection performance will inevitably be worse in the mid frequency region. However, an excessive peak in the sensitivity function is in general not desired in motion systems. As such, a limit on the peak of the sensitivity function should be set, typically 6 dB-10 dB, where a smaller value corresponds to a more robust but potentially conservative design. In our work, we set a limit of 7 dB for the peak of sensitivity function $\max|S| \leq 7$ dB, and whenever this is violated, we reduce the step size by half to damp the convergence. With these conditions, the optimization algorithm can be carried out to determine the parameters ρ_1 and ρ_2.

Remark 7.1 *The optimization problem in* (7.12) *is non-convex in general. With the gradient-based optimization, we can expect no more than a local minimum, so a reasonably good choice of the initial parameter* ρ *is necessary, instead of a random choice. In this chapter, the traditional Butterworth design is used to select the initial parameter value.*

7.3.2 Experiments Required

To make the procedures clearer, the experiments required for the sensitivity shaping are listed. The bold right superscript refers to the experiment index within a single iteration except for Experiment 0, which is only executed once at the beginning of the sensitivity shaping process. The artificially injected white noise is denoted by N while the noises that naturally exist in the real motion system are denoted by n_i.

- Experiment 0: Artificially inject white noise N from point n in Figure 7.2.

$$r^{\mathbf{0}} = 0, d^{\mathbf{0}} = 0, n^{\mathbf{0}} = N, \tag{7.22}$$

$$y^{\mathbf{0}} = H_{yn} \cdot N + H_{yn} \cdot {n_i}^{\mathbf{0}}. \tag{7.23}$$

- Experiment 1: Artificially inject white noise N from point d in Figure 7.2 and measure the output $y^{\mathbf{1}}$.

$$r^{\mathbf{1}} = 0, d^{\mathbf{1}} = N, n^{\mathbf{1}} = 0, \tag{7.24}$$

$$y^{\mathbf{1}} = H_{yd} \cdot N + H_{yn} \cdot {n_i}^{\mathbf{1}}. \tag{7.25}$$

- Experiment 1R: This is a repeated Experiment 1, which is necessary to guarantee the unbiasedness of the gradient estimation.

$$r^{\mathbf{1R}} = 0, d^{\mathbf{1R}} = N, n^{\mathbf{1R}} = 0, \tag{7.26}$$

$$y^{\mathbf{1R}} = H_{yd} \cdot N + H_{yn} \cdot {n_i}^{\mathbf{1R}}. \tag{7.27}$$

- Experiment 2: Artificially inject white noise N from point n in Figure 7.2 and measure the output $y_{\mathbf{2}}$.

$$r^{\mathbf{2}} = 0, d^{\mathbf{2}} = 0, n^{\mathbf{2}} = N, \tag{7.28}$$

$$y^{\mathbf{2}} = H_{yn} \cdot N + H_{yn} \cdot {n_i}^{\mathbf{2}}. \tag{7.29}$$

- Experiment 3: Artificially inject white noise N from point d in Figure 7.2 and measure $u_{\mathbf{3}}$.

$$r^{\mathbf{3}} = 0, d^{\mathbf{3}} = N, n^{\mathbf{3}} = 0, \tag{7.30}$$

$$u^{\mathbf{3}} = H_{ud} \cdot N + H_{ud} \cdot {n_i}^{\mathbf{3}}. \tag{7.31}$$

7.3.3 Unbiased Cost Function Gradient Estimation

It is worth mentioning that in the calculation of the cost function gradient (7.14), $y^{\mathbf{1R}}$ should be used instead of $y^{\mathbf{1}}$. The gradient of the cost function (7.14) is estimated using closed-loop experiment data, so measurement noises that inevitably exist in the actual motion system can potentially cause some errors during this estimation. For this stochastic approximation method to work, the estimation has to be unbiased with the measurement noise

$$E[\text{est}(\nabla J(\rho))] = \nabla J(\rho). \tag{7.32}$$

To prove the unbiasedness, we have the following assumptions.

Assumption 7.1 *The noises in the different experiments are independent of each other.*

Assumption 7.2 *The noise is a zero-mean, weakly stationary random variable.*

Then, we have the following theorem.

Theorem 7.1 *For the sensitivity shaping of the DOB-based motion control system as shown in Figure 7.2, under Assumption 7.1 and Assumption 7.2, the estimation of gradient of cost function* (7.11) *is unbiased under the effect of the measurement noise.*

Proof of Theorem 7.1: Following (7.17) and (7.21), the estimation of the gradient is given by

$$\begin{aligned} \text{est}(\nabla y_1^T(\rho)) &= -\frac{1}{1-Q({}^i\rho)} \cdot \frac{\partial Q({}^i\rho)}{\partial \rho} \cdot {y^{\mathbf{1}}}^T({}^i\rho) \\ &= \nabla y_1^T(\rho) + w^T, \end{aligned} \tag{7.33}$$

$$\begin{aligned} \text{est}(\nabla y_2^T(\rho)) &= \frac{1}{1-Q({}^i\rho)} \cdot \frac{\partial Q({}^i\rho)}{\partial \rho} \cdot {u^{\mathbf{3}}}^T({}^i\rho) \\ &= \nabla y_2^T(\rho) + v^T, \end{aligned} \tag{7.34}$$

where

$$w = -\frac{1}{1-Q({}^i\rho)} \cdot \frac{\partial Q({}^i\rho)}{\partial \rho} \cdot H_{yn} \cdot {n_i}^{\mathbf{1}}, \tag{7.35}$$

$$v = \frac{1}{1-Q({}^i\rho)} \cdot \frac{\partial Q({}^i\rho)}{\partial \rho} \cdot H_{ud} \cdot {n_i}^{\mathbf{3}}. \tag{7.36}$$

Since w contains noises in Experiment 1 and $y^{\mathbf{1R}}$ contains only the noises in Experiment 1R, with Assumption 7.1 and Assumption 7.2, we have

$$E[w^T \cdot y^{\mathbf{1R}}] = E[w^T] \cdot E[y^{\mathbf{1R}}], \tag{7.37}$$

$$E[w^T] = 0. \tag{7.38}$$

Similarly, since v contains noises in Experiment 3 and $y^{\mathbf{2}} - y^{\mathbf{0}}$ contains only the noises from Experiment 0 and 2, with Assumption 7.1 and Assumption 7.2, we have

$$E[v^T \cdot (y^{\mathbf{2}} - y^{\mathbf{0}})] = E[v^T] \cdot E[y^{\mathbf{2}} - y^{\mathbf{0}}], \tag{7.39}$$

$$E[v^T] = 0. \tag{7.40}$$

With (7.14), (7.33) and (7.34), the expectation of the cost function gradient estimate is then given by

$$\begin{aligned} E[\mathrm{est}(\nabla J(\rho))] &= 2w_1 L_L L_L E[\mathrm{est}(\nabla y_1^T(\rho)) y^{\mathbf{1R}}(\rho)] \\ &\quad + 2w_2 L_H L_H E[\mathrm{est}(\nabla y_2^T(\rho))(y^{\mathbf{2}}(\rho) - y^{\mathbf{0}})] \\ &= 2w_1 L_L L_L E(\nabla y_1^T(\rho) y^{\mathbf{1R}}(\rho)) + 2w_1 L_L L_L E(w^T \cdot y^{\mathbf{1R}}) \\ &\quad + 2w_2 L_H L_H E[\nabla y_2^T(\rho)(y^{\mathbf{2}}(\rho) - y^{\mathbf{0}})] \\ &\quad + 2w_2 L_H L_H E[v^T \cdot (y^{\mathbf{2}}(\rho) - y^{\mathbf{0}})] \\ &= \nabla J(\rho) + 2w_1 L_L L_L E(w^T \cdot y^{\mathbf{1R}}) \\ &\quad + 2w_2 L_H L_H E[v^T \cdot (y^{\mathbf{2}}(\rho) - y^{\mathbf{0}})]. \end{aligned} \tag{7.41}$$

With (7.37), (7.38), (7.39) and (7.40), the expectation of the cost function gradient estimate is further derived as

$$\begin{aligned} E[\mathrm{est}(\nabla J(\rho))] &= \nabla J(\rho) + 2w_1 L_L L_L E(w^T) \cdot E(y^{\mathbf{1R}}) \\ &\quad + 2w_2 L_H L_H E(v^T) \cdot E[y^{\mathbf{2}}(\rho) - y^{\mathbf{0}})] \\ &= \nabla J(\rho) + 2w_1 L_L L_L \cdot 0 \cdot E(y^{\mathbf{1R}}) \\ &\quad + 2w_2 L_H L_H \cdot 0 \cdot E[y^{\mathbf{2}}(\rho) - y^{\mathbf{0}})] \\ &= \nabla J(\rho). \end{aligned} \tag{7.42}$$

This proves Theorem 7.1 by (7.32).

This is exactly the reason why Experiment 1R is needed. If the $y^{\mathbf{1}}$ is used in the cost function gradient estimation instead of $y^{\mathbf{1R}}$, the same noise from Experiment 1 would exist in both $\mathrm{est}(\nabla y_1^T(\rho))$ and $y^{\mathbf{1}}$. Thus, $E[w^T \cdot y^{\mathbf{1}}] \neq 0$, and it leads to a biased cost function gradient estimation. An overview of the proposed sensitivity shaping optimization is illustrated in Figure 7.3.

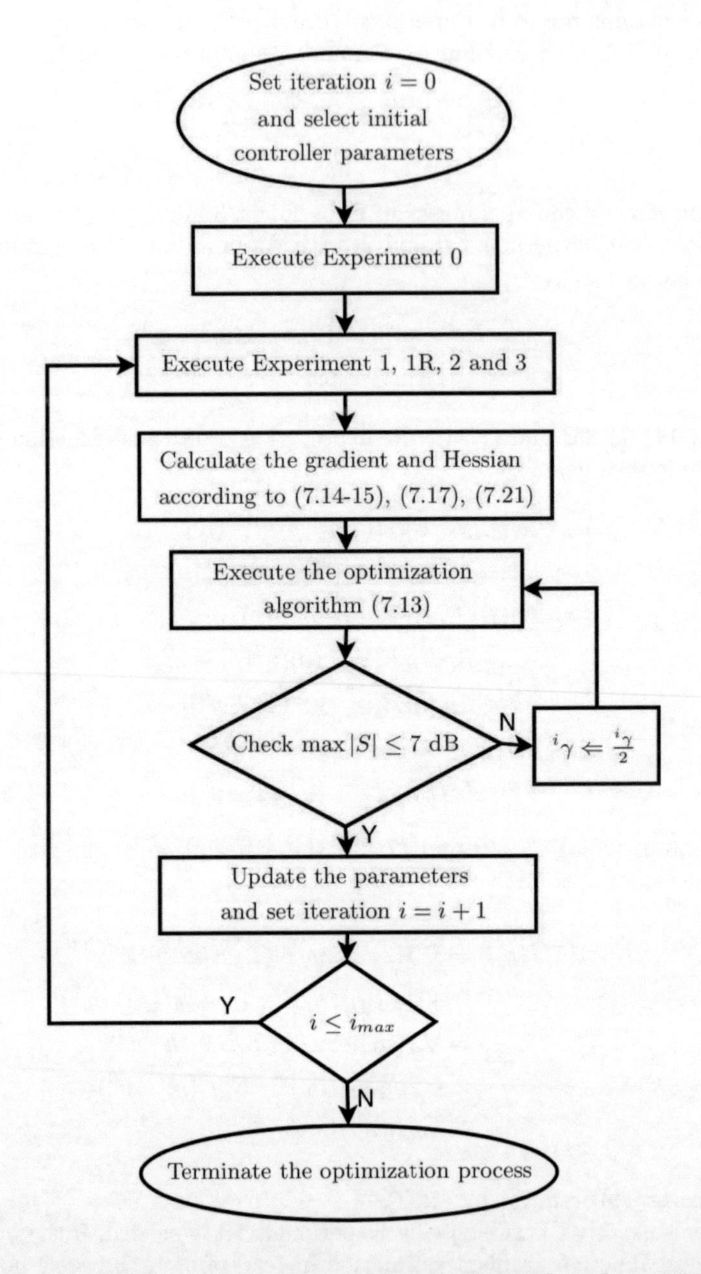

Figure 7.3: Flowchart of the proposed iterative data-based sensitivity shaping approach. © [2019] Elsevier. Reprinted, with permission, from X. Li, S.-L. Chen, C. S. Teo, and K. K. Tan, Enhanced Sensitivity Shaping by Data-Based Tuning of Disturbance Observer with Non-Binomial Filter, ISA Transactions, vol. 85, pp. 284–292, 2019.

7.4 Simulation Analysis

The simulation is conducted based on the same timing belt motion system as in Chapter 5 and Chapter 6. It is assumed that m_t can be increased to be as much as 130% of the original mass to cater to additional mass added to the system. Also, assume the first resonance frequency varies between 90% and 110% of the original frequency to take into account the position-dependent and time-dependent changes of natural frequency. Figure 7.4 shows the full-order plant $P(s)$ with its uncertainties. Consequently, the robust stability condition (7.9) has to be satisfied under such uncertainties. To meet the robust stability condition, we have to keep $T(j\omega) < \Delta(j\omega)^{-1}$ at all frequencies. An initial design of feedback controller C and the low pass filter Q has to meet this requirement for guaranteed robust stability. Thus, the feedback controller is designed as

$$C = 10 + \frac{260s}{s + 130}, \tag{7.43}$$

and the initial Q-filter is designed as

$$Q = \frac{100}{s^2 + 141s + 10000}. \tag{7.44}$$

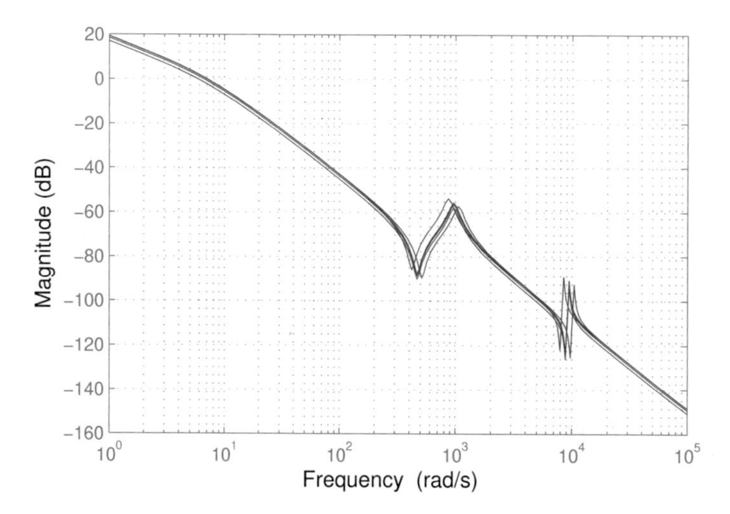

Figure 7.4: Full-order plant with perturbations. © [2019] Elsevier. Reprinted, with permission, from X. Li, S.-L. Chen, C. S. Teo, and K. K. Tan, Enhanced Sensitivity Shaping by Data-Based Tuning of Disturbance Observer with Non-Binomial Filter, ISA Transactions, vol. 85, pp. 284–292, 2019.

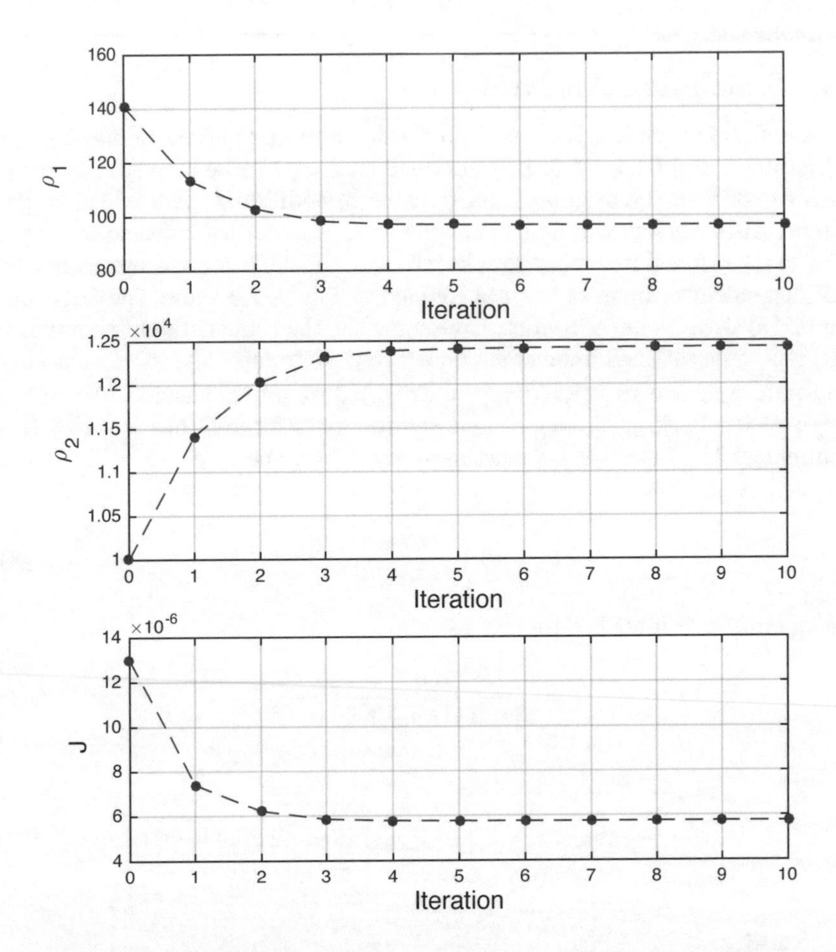

Figure 7.5: Convergence of the parameters ρ_1, ρ_2 and the cost function value. © [2019] Elsevier. Reprinted, with permission, from X. Li, S.-L. Chen, C. S. Teo, and K. K. Tan, Enhanced Sensitivity Shaping by Data-Based Tuning of Disturbance Observer with Non-Binomial Filter, ISA Transactions, vol. 85, pp. 284–292, 2019.

with a cut-off frequency at 100 rad/s. The initial complementary sensitivity function T as well as the inverse of a set of uncertainties $\Delta(j\omega)^{-1}$ are plotted in Figure 7.6 and we can notice that the robust stability condition is satisfied because $T(j\omega)$ is below $\Delta(j\omega)^{-1}$ for all frequencies. After a few iterations of tuning, the optimized S and T are plotted against the original ones in Figure 7.6, where the original ones are plotted using dashed lines. We can observe that S has been reduced in the low frequency region, while T in the high frequency region has no significant changes (overlap with each other in the figure); thus the robust stability condition is still satisfied. The control

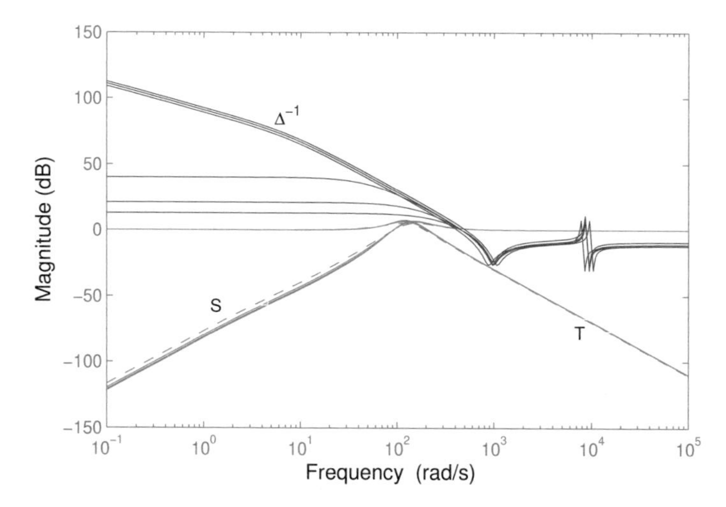

Figure 7.6: Change of S and T as well as inverse of uncertainty. © [2019] Elsevier. Reprinted, with permission, from X. Li, S.-L. Chen, C. S. Teo, and K. K. Tan, Enhanced Sensitivity Shaping by Data-Based Tuning of Disturbance Observer with Non-Binomial Filter, ISA Transactions, vol. 85, pp. 284–292, 2019.

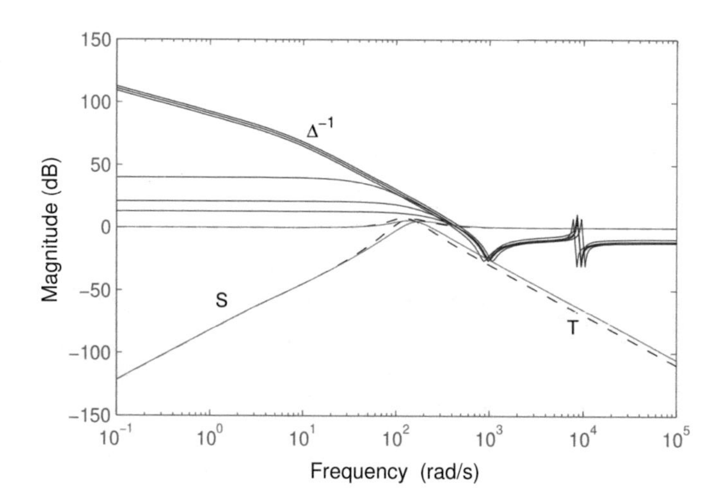

Figure 7.7: Comparison with the traditional Q-filter design with similar low frequency disturbance rejection performance. Dashed lines: Achieved S and T with data-based optimization. Solid lines: S and T with traditional design. Solid dark lines: inverse of uncertainty. © [2019] Elsevier. Reprinted, with permission, from X. Li, S.-L. Chen, C. S. Teo, and K. K. Tan, Enhanced Sensitivity Shaping by Data-Based Tuning of Disturbance Observer with Non-Binomial Filter, ISA Transactions, vol. 85, pp. 284–292, 2019.

objective to improve disturbance rejection performance while not sacrificing robust stability and high frequency noise attenuation is achieved. In contrast, this is not achievable with the traditional design because any attempt to improve the disturbance rejection performance by extending the bandwidth will result in an increase of T in the high frequency region. As shown in Figure 7.7, if similar disturbance rejection performance is achieved by using the traditional Butterworth form, the robust stability condition is violated at around 1000 rad/s as the solid line crosses the inverse of the uncertainty plot. The change of parameters ρ_1, ρ_2 and the cost function value are plotted in Figure 7.5. The final Q-filter is given by

$$Q = \frac{12424}{s^2 + 96.58s + 12424}. \tag{7.45}$$

Figure 7.8 shows the time domain step and sinusoidal wave disturbance rejection performance. The black line shows the output response to

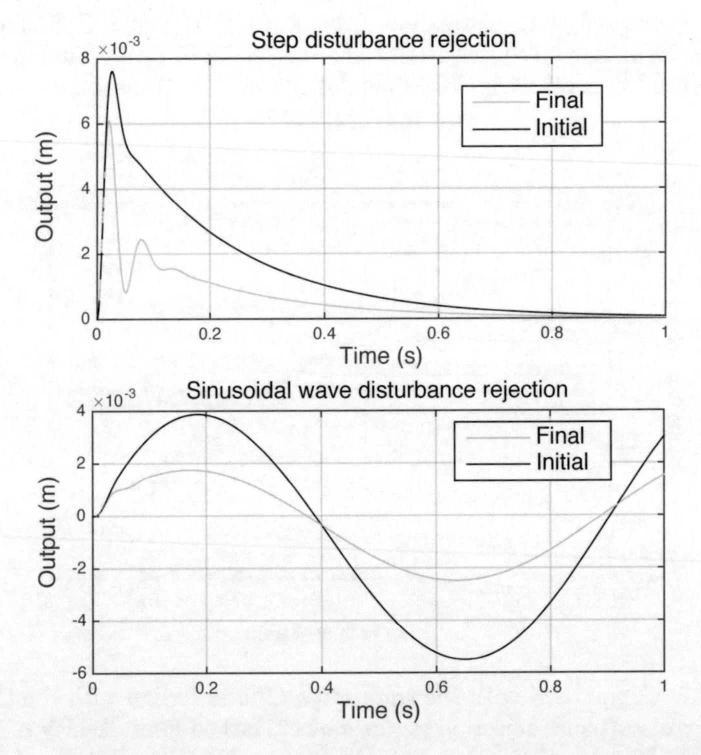

Figure 7.8: Time domain disturbance rejection performance. © [2019] Elsevier. Reprinted, with permission, from X. Li, S.-L. Chen, C. S. Teo, and K. K. Tan, Enhanced Sensitivity Shaping by Data-Based Tuning of Disturbance Observer with Non-Binomial Filter, ISA Transactions, vol. 85, pp. 284–292, 2019.

disturbances with a traditional Q-filter and the grey line shows the response with the optimized Q-filter, and the improvement can be observed.

Apart from the 2nd-order Q-filter which is most commonly used in the industrial application, the proposed method is also applicable to higher-order Q-filters. To further demonstrate the effectiveness of the proposed method, an additional simulation case study based on a 4th-order Q-filter is conducted. Similarly, the traditional Butterworth design is used to initialize the Q-filter in the first iteration, where

$$Q = \frac{100000000}{s^4 + 261.31s^3 + 34142s^2 + 2613100s + 100000000}. \tag{7.46}$$

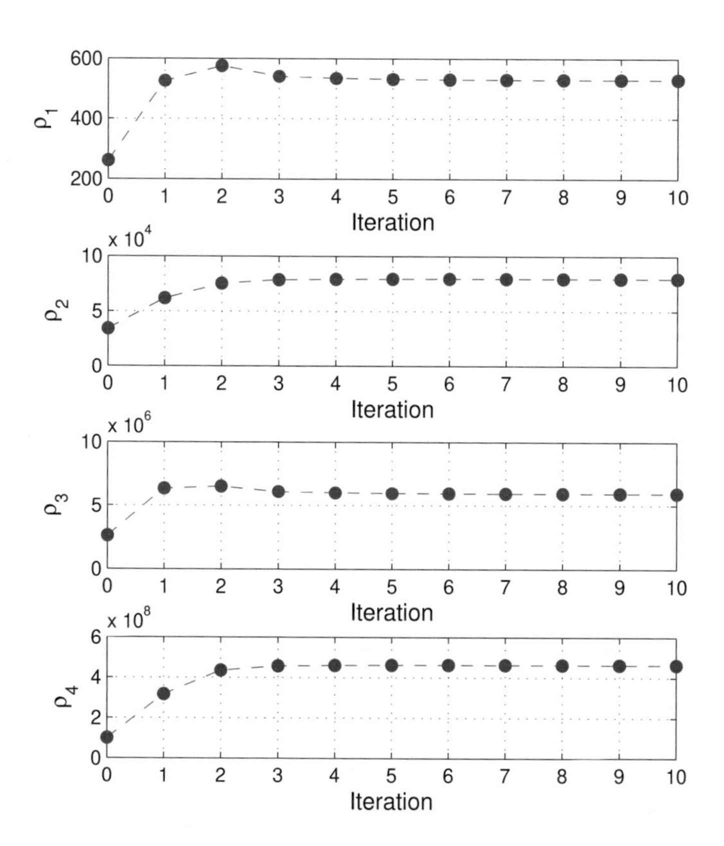

Figure 7.9: Convergence of the parameters ρ_1, ρ_2, ρ_3 and ρ_4 for 4th-order Q-filter. © [2019] Elsevier. Reprinted, with permission, from X. Li, S.-L. Chen, C. S. Teo, and K. K. Tan, Enhanced Sensitivity Shaping by Data-Based Tuning of Disturbance Observer with Non-Binomial Filter, ISA Transactions, vol. 85, pp. 284–292, 2019.

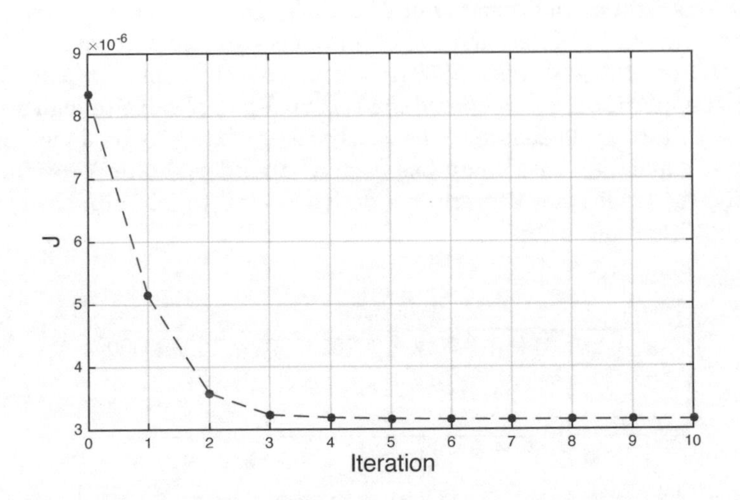

Figure 7.10: Cost function reduction for 4th-order Q-filter from iteration 1 to 10. © [2019] Elsevier. Reprinted, with permission, from X. Li, S.-L. Chen, C. S. Teo, and K. K. Tan, Enhanced Sensitivity Shaping by Data-Based Tuning of Disturbance Observer with Non-Binomial Filter, ISA Transactions, vol. 85, pp. 284–292, 2019.

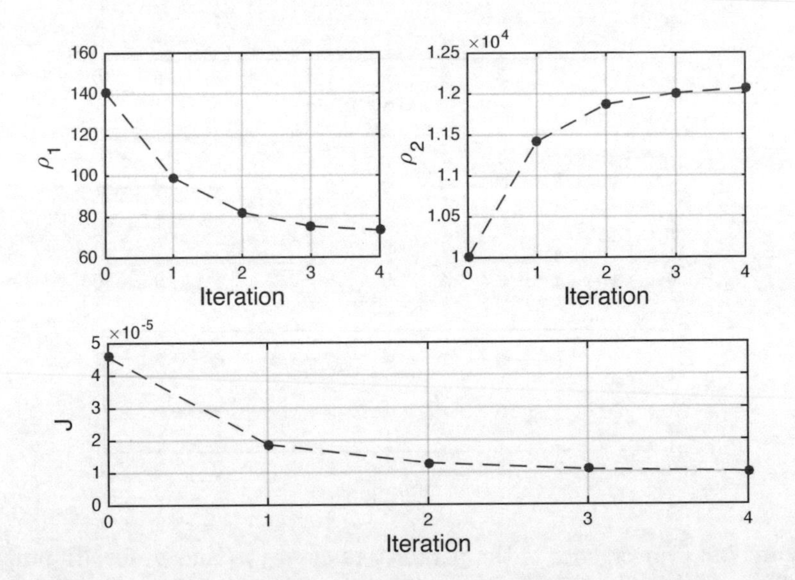

Figure 7.11: Convergence of Q-filter parameters and reduction of cost function value in the experiment. © [2019] Elsevier. Reprinted, with permission, from X. Li, S.-L. Chen, C. S. Teo, and K. K. Tan, Enhanced Sensitivity Shaping by Data-Based Tuning of Disturbance Observer with Non-Binomial Filter, ISA Transactions, vol. 85, pp. 284–292, 2019.

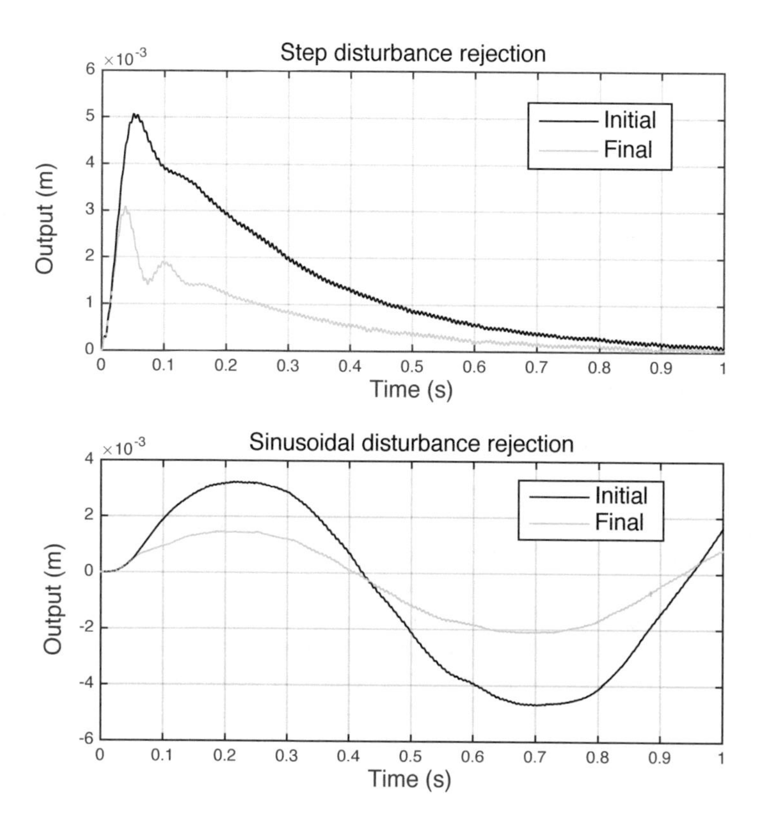

Figure 7.12: Improved disturbance rejection performance in the experiment. © [2019] Elsevier. Reprinted, with permission, from X. Li, S.-L. Chen, C. S. Teo, and K. K. Tan, Enhanced Sensitivity Shaping by Data-Based Tuning of Disturbance Observer with Non-Binomial Filter, ISA Transactions, vol. 85, pp. 284–292, 2019.

With the proposed data-based sensitivity shaping approach, Figure 7.9 shows the convergence of the four parameters in 10 iterations and a significant reduction of the cost function is shown in Figure 7.10. The final Q-filter is given by

$$Q = \frac{4.62 \times 10^8}{s^4 + 528.88s^3 + 7.95 \times 10^4 s^2 + 5.93 \times 10^6 s + 4.62 \times 10^8}. \tag{7.47}$$

7.5 Experimental Validation

In this section, the proposed data-based sensitivity shaping optimization is verified on the timing belt setup. By executing the proposed data-based optimization procedures, the parameter evolution, as well as the cost function reduction, are shown in Figure 7.11. Due to the uncertainties that exist in the practical situation, the algorithm is stopped after 4 iterations, which shows a reasonably good result. Further optimization will only result in fluctuations around the optimal values and is less meaningful in industrial applications. Figure 7.12 shows the system output when the step disturbance and the sinusoidal disturbance are injected, and significant improvement can be observed with the optimized Q-filter.

7.6 Conclusion

In this chapter, a data-based optimization approach for DOB sensitivity shaping is proposed. Instead of the traditional design, where the only parameter is the bandwidth, we make the Q-filter a more general form with additional freedom so that its sensitivity shaping is more flexible. The parameters of the Q-filter are optimized using a data-based approach, and the control objective is to improve low frequency disturbance rejection while not deteriorating the high frequency noise attenuation performance as well as the robust stability. The effectiveness of the proposed method is illustrated using both simulations and experiments, and the improvement is clearly shown in both frequency and time domains. Similar performance is not achievable with the traditional design because any attempt to reduce the low frequency sensitivity function will result in rising of the complementary sensitivity function in the high frequency region and consequently violates the robust stability criterion.

Bibliography

Al Mamun, A., G. Guo, and C. Bi (2006). *Hard Disk Drive: Mechatronics and Control.* Boca Raton: CRC Press.

An, H., J. Liu, C. Wang, and L. Wu (2016). Disturbance observer-based antiwindup control for air-breathing hypersonic vehicles. *IEEE Transactions on Industrial Electronics 63*(5), 3038–3049.

Anderson, B. D. O. and J. B. Moore (1971). *Linear Optimal Control.* Englewood Cliffs: Prentice-Hall.

Anderson, B. D. O. and J. B. Moore (2007). *Optimal Control: Linear Quadratic Methods.* Mineola: Dover Publications.

Åström, K. J. and T. Hägglund (1995). *PID Controllers: Theory, Design, and Tuning.* Research Triangle Park: International Society of Automation.

Åström, K. J. and B. Wittenmark (2013). *Adaptive Control.* Mineola: Dover Publications.

Bakule, L. (2008). Decentralized control: an overview. *Annual Reviews in Control 32*(1), 87–98.

Barmish, B. R. (1983). Stabilization of uncertain systems via linear control. *IEEE Transactions on Automatic Control 28*(8), 848–850.

Barmish, B. R. (1985). Necessary and sufficient conditions for quadratic stabilizability of an uncertain system. *Journal of Optimization Theory and Applications 46*(4), 399–408.

Bazaraa, M. S., H. D. Sherali, and C. M. Shetty (2005). *Nonlinear Programming: Theory and Algorithms.* New York: John Wiley & Sons.

Becerra, V. M. (2004). Solving optimal control problems with state constraints using nonlinear programming and simulation tools. *IEEE Transactions on Education 47*(3), 377–384.

Bellman, R. (1957). *Dynamic Programming.* Princeton: Princeton University Press.

Bernstein, D. S. and W. M. Haddad (1989). LQG control with an H_∞ performance bound: a Riccati equation approach. *IEEE Transactions on Automatic Control 34*(3), 293–305.

Bernussou, J., P. L. D. Peres, and J. C. Geromel (1989). A linear programming oriented procedure for quadratic stabilization of uncertain systems. *Systems & Control Letters 13*(1), 65–72.

Bertsekas, D. P. (1995). *Dynamic Programming and Optimal Control.* Belmont: Athena Scientific.

Bertsekas, D. P. (1999). *Nonlinear Programming.* Belmont: Athena Scientific.

Bertsekas, D. P. and H. Yu (2011). A unifying polyhedral approximation framework for convex optimization. *SIAM Journal on Optimization 21*(1), 333–360.

Betts, J. T. (2001). *Practical Methods for Optimal Control Using Nonlinear Programming.* Warrendale: SIAM.

Bhattacharyya, S. P. and L. H. Keel (1994). Robust control: the parametric approach. *IFAC Proceedings Volumes 27*(9), 49–52.

Bishop, R. H. (2005). *Mechatronics: an Introduction.* Boca Raton: CRC Press.

Boender, C. G. E., A. R. Kan, G. Timmer, and L. Stougie (1982). A stochastic method for global optimization. *Mathematical Programming 22*(1), 125–140.

Boerlage, M., R. Tousain, and M. Steinbuch (2004). Jerk derivative feedforward control for motion systems. In *Proceedings of 2004 American Control Conference*, Volume 5, pp. 4843–4848.

Böhme, T. J. and B. Frank (2017). *Hybrid Systems, Optimal Control and Hybrid Vehicles: Theory, Methods and Applications.* Cham: Springer.

Boyd, S. and L. Vandenberghe (2004). *Convex Optimization.* Cambridge: Cambridge University Press.

Bozca, M., A. Muğan, and H. Temeltaş (2008). Decoupled approach to integrated optimum design of structures and robust control systems. *Structural and Multidisciplinary Optimization 36*(2), 169–191.

Bristow, D. A., M. Tharayil, and A. G. Alleyne (2006). A survey of iterative learning control. *IEEE Control Systems 26*(3), 96–114.

Burl, J. B. (1999). *Linear Optimal Control: H_2 and H_∞ Methods.* Menlo Park: Addison-Wesley.

Byers, R. (1988). A bisection method for measuring the distance of a stable matrix to the unstable matrices. *SIAM Journal on Scientific and Statistical Computing 9*(5), 875–881.

Campanelli, S., G. Cardano, R. Giannoccaro, A. Ludovico, and E. L. Bohez (2007). Statistical analysis of the stereolithographic process to improve the accuracy. *Computer-Aided Design 39*(1), 80–86.

Campi, M. C. and S. M. Savaresi (2006). Direct nonlinear control design: the virtual reference feedback tuning (VRFT) approach. *IEEE Transactions on Automatic Control 51*(1), 14–27.

Castro-Garcia, R., K. Tiels, O. M. Agudelo, and J. A. K. Suykens (2018). Hammerstein system identification through best linear approximation inversion and regularisation. *International Journal of Control 91*(8), 1757–1773.

Chen, B. M. and A. Saberi (1993). Necessary and sufficient conditions under which an H_2 optimal control problem has a unique solution. *International Journal of Control 58*(2), 337–348.

Chen, B. M., A. Saberi, P. Sannuti, and Y. Shamash (1993). Construction and parameterization of all static and dynamic H_2-optimal state feedback solutions, optimal fixed modes and fixed decoupling zeros. *IEEE Transactions on Automatic Control 38*(2), 248–261.

Chen, S.-L., N. Kamaldin, T. J. Teo, W. Liang, C. S. Teo, G. Yang, and K. K. Tan (2015). Toward comprehensive modeling and large-angle tracking control of a limited-angle torque actuator with cylindrical halbach. *IEEE/ASME Transactions on Mechatronics 21*(1), 431–442.

Chen, S.-L., X. Li, C. S. Teo, and K. K. Tan (2017). Composite jerk feedforward and disturbance observer for robust tracking of flexible systems. *Automatica 80*, 253–260.

Chen, S.-L., K. K. Tan, and S. Huang (2012). Identification of coulomb friction-impeded systems with a triple-relay feedback apparatus. *IEEE Transactions on Control Systems Technology 20*(3), 726–737.

Chen, S.-L., K. K. Tan, S. Huang, and C. S. Teo (2010). Modeling and compensation of ripples and friction in permanent-magnet linear motor using a hysteretic relay. *IEEE/ASME Transactions on Mechatronics 15*(4), 586–594.

Chen, X., L. Wu, Y. Deng, and Q. Wang (2017). Dynamic response analysis and chaos identification of 4-UPS-UPU flexible spatial parallel mechanism. *Nonlinear Dynamics 87*(4), 2311–2324.

Chen, Z., B. Yao, and Q. Wang (2013). Adaptive robust precision motion control of linear motors with integrated compensation of nonlinearities and bearing flexible modes. *IEEE Transactions on Industrial Informatics 9*(2), 965–973.

Chen, Z., B. Yao, and Q. Wang (2015). μ-synthesis-based adaptive robust control of linear motor driven stages with high-frequency dynamics: A case study. *IEEE/ASME Transactions on Mechatronics 20*(3), 1482–1490.

Chevalier, A., C. Copot, C. M. Ionescu, and R. De Keyser (2016). Automatic calibration with robust control of a six DOF mechatronic system. *Mechatronics 35*, 102–108.

Chi, R., Y. Liu, Z. Hou, and S. Jin (2015). Data-driven terminal iterative learning control with high-order learning law for a class of non-linear discrete-time multiple-input–multiple output systems. *IET Control Theory & Applications 9*(7), 1075–1082.

Cho, H., W. Park, B. Choi, and M.-C. Leu (2000). Determining optimal parameters for stereolithography processes via genetic algorithm. *Journal of Manufacturing Systems 19*(1), 18–27.

Choi, Y., K. Yang, W. K. Chung, H. R. Kim, and I. H. Suh (2003). On the robustness and performance of disturbance observers for second-order systems. *IEEE Transactions on Automatic Control 48*(2), 315–320.

Clayton, G. M., S. Tien, K. K. Leang, Q. Zou, and S. Devasia (2009). A review of feedforward control approaches in nanopositioning for high-speed SPM. *Journal of Dynamic Systems, Measurement, and Control 131*(6), 061101.

Corana, A., M. Marchesi, C. Martini, and S. Ridella (1987). Minimizing multimodal functions of continuous variables with the simulated annealing algorithm. *ACM Transactions on Mathematical Software 13*(3), 262–280.

Cordeau, J.-F., M. Gendreau, and G. Laporte (1997). A tabu search heuristic for periodic and multi-depot vehicle routing problems. *Networks 30*(2), 105–119.

Cvejn, J. and J. Tvrdík (2017). Learning control of a robot manipulator based on a decentralized position-dependent pid controller. In *Proceedings of the 21st International Conference on Process Control*, pp. 167–172.

Dantzig, G. (2016). *Linear Programming and Extensions.* Princeton: Princeton University Press.

Davidon, W. C. (1991). Variable metric method for minimization. *SIAM Journal on Optimization 1*(1), 1–17.

Deb, K., A. Pratap, S. Agarwal, and T. Meyarivan (2002). A fast and elitist multiobjective genetic algorithm: NSGA-II. *IEEE Transactions on Evolutionary Computation 6*(2), 182–197.

Dolk, V. S., D. P. Borgers, and W. P. M. H. Heemels (2017). Output-based and decentralized dynamic event-triggered control with guaranteed $\mathcal{L}_p$-gain performance and zeno-freeness. *IEEE Transactions on Automatic Control 62*(1), 34–49.

Doyle, J. C., B. A. Francis, and A. R. Tannenbaum (1992). *Feedback Control Theory.* New York: Maxwell MacMillan International.

Doyle, J. C., K. Glover, P. P. Khargonekar, and B. A. Francis (1989). State-space solutions to standard H_2 and H_∞ control problems. *IEEE Transactions on Automatic Control 34*(8), 831–847.

Epperson, J. F. (1987). On the runge example. *The American Mathematical Monthly 94*(4), 329–341.

Farjood, E., M. Vojdani, K. Torabi, and A. A. R. Khaledi (2017). Marginal and internal fit of metal copings fabricated with rapid prototyping and conventional waxing. *The Journal of Prosthetic Dentistry 117*(1), 164–170.

Fattahi, S. and J. Lavaei (2017). On the convexity of optimal decentralized control problem and sparsity path. In *Proceedings of the 2017 American Control Conference*, pp. 3359–3366.

Formentin, S. and A. Karimi (2013). A data-driven approach to mixed-sensitivity control with application to an active suspension system. *IEEE Transactions on Industrial Informatics 9*(4), 2293–2300.

García-Herreros, I., X. Kestelyn, J. Gomand, R. Coleman, and P.-J. Barre (2013). Model-based decoupling control method for dual-drive gantry stages: A case study with experimental validations. *Control Engineering Practice 21*(3), 298–307.

Geromel, J. C. (1999). Optimal linear filtering under parameter uncertainty. *IEEE Transactions on Signal Processing 47*(1), 168–175.

Geromel, J. C. (University Paul Sabatier, 1979). *Contribution à l'étude des systèmes dynamiques interconnectes: aspects de decentralisation.* Ph. D. thesis.

Geromel, J. C. and J. Bernussou (1982). Optimal decentralized control of dynamic systems. *Automatica 18*(5), 545–557.

Geromel, J. C., J. Bernussou, and P. L. D. Peres (1994). Decentralized control through parameter space optimization. *Automatica 30*(10), 1565–1578.

Geromel, J. C., P. L. D. Peres, and J. Bernussou (1991). On a convex parameter space method for linear control design of uncertain systems. *SIAM Journal on Control and Optimization 29*(2), 381–402.

Geromel, J. C., P. L. D. Peres, and S. R. Souza (1992). H_2 guaranteed cost control for uncertain continuous-time linear systems. *Systems & Control Letters 19*(1), 23–27.

Geromel, J. C., P. L. D. Peres, and S. R. Souza (1993). Convex analysis of output feedback structural constraints. In *Proceedings of the 32nd IEEE Conference on Decision and Control*, pp. 1363–1364.

Giam, T. S., K. K. Tan, and S. Huang (2007). Precision coordinated control of multi-axis gantry stages. *ISA Transactions 46*(3), 399–409.

Goldstine, H. H. (1981). *A History of the Calculus of Variations from the 17th through the 19th Century.* New York: Springer-Verlag.

Han, Q.-L. (2004). On robust stability of neutral systems with time-varying discrete delay and norm-bounded uncertainty. *Automatica 40*(6), 1087–1092.

Hao, R., J. Wang, J. Zhao, and S. Wang (2016). Observer-based robust control of 6-DOF parallel electrical manipulator with fast friction estimation. *IEEE Transactions on Automation Science and Engineering 13*(3), 1399–1408.

Heertjes, M. F. (2016). Data-based motion control of wafer scanners. *IFAC-PapersOnLine 49*(13), 1–12.

Hjalmarsson, H. (1998). Control of nonlinear systems using iterative feedback tuning. In *Proceedings of the 1998 American Control Conference*, Volume 4, pp. 2083–2087.

Hjalmarsson, H. (2002). Iterative feedback tuning-an overview. *International Journal of Adaptive Control and Signal Processing 16*(5), 373–395.

Hjalmarsson, H., M. Gevers, S. Gunnarsson, and O. Lequin (1998). Iterative feedback tuning: theory and applications. *IEEE Control Systems 18*(4), 26–41.

Hjalmarsson, H., S. Gunnarsson, and M. Gevers (1994). A convergent iterative restricted complexity control design scheme. In *Proceedings of the 33rd IEEE Conference on Decision and Control*, Volume 2, pp. 1735–1740.

Hou, H., X. Nian, H. Xiong, Z. Wang, and Z. Peng (2016). Robust decentralized coordinated control of a multimotor web-winding system. *IEEE Transactions on Control Systems Technology 24*(4), 1495–1503.

Hou, Z. and S. Jin (2011a). Data-driven model-free adaptive control for a class of MIMO nonlinear discrete-time systems. *IEEE Transactions on Neural Networks 22*(12), 2173–2188.

Hou, Z. and S. Jin (2011b). A novel data-driven control approach for a class of discrete-time nonlinear systems. *IEEE Transactions on Control Systems Technology 19*(6), 1549–1558.

Hou, Z. and Z. Wang (2013). From model-based control to data-driven control: Survey, classification and perspective. *Information Sciences 235*, 3–35.

Hou, Z. and J.-X. Xu (2009). On data-driven control theory: the state of the art and perspective. *Acta Automatica Sinica 35*(6), 650–667.

Hu, C., Z. Hu, Y. Zhu, and Z. Wang (2017). Advanced GTCF-LARC contouring motion controller design for an industrial X-Y linear motor stage with experimental investigation. *IEEE Transactions on Industrial Electronics 64*(4), 3308–3318.

Hu, C., Z. Wang, Y. Zhu, M. Zhang, and H. Liu (2016). Performance-oriented precision LARC tracking motion control of a magnetically levitated planar motor with comparative experiments. *IEEE Transactions on Industrial Electronics 63*(9), 5763–5773.

Hu, C., B. Yao, and Q. Wang (2010). Coordinated adaptive robust contouring control of an industrial biaxial precision gantry with cogging force compensations. *IEEE Transactions on Industrial Electronics 57*(5), 1746–1754.

Hu, C., B. Yao, and Q. Wang (2011). Global task coordinate frame-based contouring control of linear-motor-driven biaxial systems with accurate parameter estimations. *IEEE Transactions on Industrial Electronics 58*(11), 5195–5205.

Jang, J. S. R., C. T. Sun, and E. Mizutani (1997). Neuro-fuzzy and soft computing-a computational approach to learning and machine intelligence. *IEEE Transactions on Automatic Control 42*(10), 1482–1484.

Jiang, Y., Y. Zhu, K. Yang, C. Hu, and D. Yu (2015). A data-driven iterative decoupling feedforward control strategy with application to an ultraprecision motion stage. *IEEE Transactions on Industrial Electronics 62*(1), 620–627.

Kammer, L. C., R. R. Bitmead, and P. L. Bartlett (2000). Direct iterative tuning via spectral analysis. *Automatica 36*(9), 1301–1307.

Karimi, A., L. Mišković, and D. Bonvin (2004). Iterative correlation-based controller tuning. *International Journal of Adaptive Control and Signal Processing 18*(8), 645–664.

Kempf, C. J. and S. Kobayashi (1999). Disturbance observer and feedforward design for a high-speed direct-drive positioning table. *IEEE Transactions on Control Systems Technology 7*(5), 513–526.

Khargonekar, P. P. and M. A. Rotea (1991). Mixed H_2/H_∞ control: a convex optimization approach. *IEEE Transactions on Automatic Control 36*(7), 824–837.

Kopp, R. E. (1962). Pontryagin maximum principle. *Mathematics in Science and Engineering 5*, 255–279.

Kučera, V. (2007). The H_2 control problem: a general transfer-function solution. *International Journal of Control 80*(5), 800–815.

Kuriyama, K., H. Onishi, N. Sano, T. Komiyama, Y. Aikawa, Y. Tateda, T. Araki, and M. Uematsu (2003). A new irradiation unit constructed of self-moving gantry-CT and linac. *International Journal of Radiation Oncology, Biology, Physics 2*(55), 428–435.

Lambrechts, P., M. Boerlage, and M. Steinbuch (2005). Trajectory planning and feedforward design for electromechanical motion systems. *Control Engineering Practice 13*(2), 145–157.

Lee, J. and P. Hajela (1996). Parallel genetic algorithm implementation in multidisciplinary rotor blade design. *Journal of Aircraft 33*(5), 962–969.

Leitmann, G. (1981). *The Calculus of Variations and Optimal Control: an Introduction.* New York: Plenum Press.

Li, C., B. Yao, X. Zhu, and Q. Wang (2015). Dual drive system modeling and analysis for synchronous control of an H-type gantry. In *Proceedings of the 2015 IEEE/ASME International Conference on Advanced Intelligent Mechatronics*, pp. 214–219.

Li, D. and S. W. Yoon (2017). PCB assembly optimization in a single gantry high-speed rotary-head collect-and-place machine. *The International Journal of Advanced Manufacturing Technology 88*(9), 2819–2834.

Li, F.-D., M. Wu, Y. He, and X. Chen (2012). Optimal control in microgrid using multi-agent reinforcement learning. *ISA Transactions 51*(6), 743–751.

Li, M., Y. Zhu, K. Yang, and C. Hu (2015). A data-driven variable-gain control strategy for an ultra-precision wafer stage with accelerated iterative parameter tuning. *IEEE Transactions on Industrial Informatics 11*(5), 1179–1189.

Li, M., Y. Zhu, K. Yang, C. Hu, and H. Mu (2017). An integrated model-data-based zero-phase error tracking feedforward control strategy with application to an ultraprecision wafer stage. *IEEE Transactions on Industrial Electronics 64*(5), 4139–4149.

Li, X. (National University of Singapore, 2017). *Data-based control design with application to tray indexing.* Ph. D. thesis.

Li, X., S.-L. Chen, C. S. Teo, and K. K. Tan (2017). Data-based tuning of reduced-order inverse model in both disturbance observer and feedforward with application to tray indexing. *IEEE Transactions on Industrial Electronics 64*(7), 5492–5501.

Li, X., S.-L. Chen, C. S. Teo, and K. K. Tan (2019). Enhanced sensitivity shaping by data-based tuning of disturbance observer with non-binomial filter. *ISA Transactions 85*, 284–292.

Li, X., S.-L. Chen, C. S. Teo, K. K. Tan, and T. H. Lee (2015). Data-driven modeling of control valve stiction using revised binary-tree structure. *Industrial & Engineering Chemistry Research 54*(1), 330–337.

Li, Z., C.-Y. Su, L. Wang, Z. Chen, and T. Chai (2015). Nonlinear disturbance observer-based control design for a robotic exoskeleton incorporating fuzzy approximation. *IEEE Transactions on Industrial Electronics 62*(9), 5763–5775.

Liu, L., K. K. Tan, S.-L. Chen, S. Huang, and T. H. Lee (2012). SVD-based preisach hysteresis identification and composite control of piezo actuators. *ISA Transactions 51*(3), 430–438.

Liu, L., K. K. Tan, S.-L. Chen, C. S. Teo, and T. H. Lee (2013). Discrete composite control of piezoelectric actuators for high-speed and precision scanning. *IEEE Transactions on Industrial Informatics 9*(2), 859–868.

Liu, L., K. K. Tan, C. S. Teo, S.-L. Chen, and T. H. Lee (2013). Development of an approach toward comprehensive identification of hysteretic dynamics in piezoelectric actuators. *IEEE Transactions on Control Systems Technology 21*(5), 1834–1845.

Lu, L., Z. Chen, B. Yao, and Q. Wang (2013). A two-loop performance-oriented tip-tracking control of a linear-motor-driven flexible beam system with experiments. *IEEE Transactions on Industrial Electronics 60*(3), 1011–1022.

Ma, J. (National University of Singapore, 2017). *Constrained optimization algorithms towards integrated design of mechatronic systems.* Ph. D. thesis.

Ma, J., S.-L. Chen, N. Kamaldin, C. S. Teo, A. Tay, A. Al Mamun, and K. K. Tan (2017). A novel constrained H_2 optimization algorithm for mechatronics design in flexure-linked biaxial gantry. *ISA Transactions 71*, 467–479.

Ma, J., S.-L. Chen, N. Kamaldin, C. S. Teo, A. Tay, A. Al Mamun, and K. K. Tan (2018). Integrated mechatronic design in the flexure-linked dual-drive gantry by constrained linear-quadratic optimization. *IEEE Transactions on Industrial Electronics 65*(3), 2408–2418.

Ma, J., S.-L. Chen, W. Liang, C. S. Teo, A. Tay, A. Al Mamun, and K. K. Tan (2019). Robust decentralized controller synthesis in flexure-linked H-gantry by iterative linear programming. *IEEE Transactions on Industrial Informatics 15*(3), 1698–1708.

Ma, J., S.-L. Chen, C. S. Teo, C. J. Kong, A. Tay, W. Lin, and A. Al Mamun (2017). A constrained linear quadratic optimization algorithm toward jerk-decoupling cartridge design. *Journal of the Franklin Institute 354*(1), 479–500.

Ma, J., S.-L. Chen, C. S. Teo, A. Tay, A. Al Mamun, and K. K. Tan (2019). Parameter space optimization towards integrated mechatronic design for uncertain systems with generalized feedback constraints. *Automatica 105*, 149–158.

Mondal, B., K. Dasgupta, and P. Dutta (2012). Load balancing in cloud computing using stochastic hill climbing-a soft computing approach. *Procedia Technology 4*, 783–789.

Nazareth, L. and P. Tseng (2002). Gilding the lily: A variant of the Nelder-Mead algorithm based on golden-section search. *Computational Optimization and Applications 22*(1), 133–144.

Nelder, J. A. and R. Mead (1965). A simplex method for function minimization. *The Computer Journal 7*(4), 308–313.

Palhares, R. M., R. H. C. Taicahashi, and P. L. D. Peres (1997). H_∞ and H_2 guaranteed costs computation for uncertain linear systems. *International Journal of Systems Science 28*(2), 183–188.

Paul, R. and S. Anand (2012). Process energy analysis and optimization in selective laser sintering. *Journal of Manufacturing Systems 31*(4), 429–437.

Peres, P. L. D., J. C. Geromel, and J. Bernussou (1993). Quadratic stabilizability of linear uncertain systems in convex-bounded domains. *Automatica 29*(2), 491–493.

Peres, P. L. D., J. C. Geromel, and S. R. Souza (1993). Optimal H_2 control by output feedback. In *Proceedings of the 32nd IEEE Conference on Decision and Control*, pp. 102–107.

Petersen, I. R. (1995). Guaranteed cost LQG control of uncertain linear systems. *IEE Proceedings-Control Theory and Applications 142*(2), 95–102.

Petersen, I. R. and C. V. Hollot (1986). A Riccati equation approach to the stabilization of uncertain linear systems. *Automatica 22*(4), 397–411.

Petersen, I. R. and D. C. McFarlane (1994). Optimal guaranteed cost control and filtering for uncertain linear systems. *IEEE Transactions on Automatic Control 39*(9), 1971–1977.

Pipelzadeh, Y., B. Chaudhuri, and T. C. Green (2013). Control coordination within a VSC HVDC link for power oscillation damping: A robust decentralized approach using homotopy. *IEEE Transactions on Control Systems Technology 21*(4), 1270–1279.

Pontryagin, L. S. (1987). *Mathematical Theory of Optimal Processes.* Boca Raton: CRC Press.

Powell, M. J. D. (1977). Restart procedures for the conjugate gradient method. *Mathematical Programming 12*(1), 241–254.

Radac, M.-B. and R.-E. Precup (2015). Data-based two-degree-of-freedom iterative control approach to constrained non-linear systems. *IET Control Theory & Applications 9*(7), 1000–1010.

Rajemi, M., P. Mativenga, and A. Aramcharoen (2010). Sustainable machining: selection of optimum turning conditions based on minimum energy considerations. *Journal of Cleaner Production 18*(10-11), 1059–1065.

Rao, A. V. (2009). A survey of numerical methods for optimal control. *Advances in the Astronautical Sciences 135*(1), 497–528.

Ren, Q., J.-X. Xu, and X. Li (2015). A data-driven motion control approach for a robotic fish. *Journal of Bionic Engineering 12*(3), 382–394.

Roberts, P. D. and V. M. Becerra (2000). Optimal control of nonlinear differential algebraic equation systems. In *Proceedings of the 39th IEEE Conference on Decision and Control*, Volume 1, pp. 754–759.

Rotondo, D., F. Nejjari, and V. Puig (2015). Robust quasi–LPV model reference FTC of a quadrotor UAV subject to actuator faults. *International Journal of Applied Mathematics and Computer Science 25*(1), 7–22.

Rădac, M. B., R. E. Precup, E. M. Petriu, S. Preitl, and C. A. Dragoş (2013). Data-driven reference trajectory tracking algorithm and experimental validation. *IEEE Transactions on Industrial Informatics 9*(4), 2327–2336.

Rugh, W. J. and J. S. Shamma (2000). Research on gain scheduling. *Automatica 36*(10), 1401–1425.

Rupp, D. and L. Guzzella (2010). Iterative tuning of internal model controllers with application to air/fuel ratio control. *IEEE Transactions on Control Systems Technology 18*(1), 177–184.

Sariyildiz, E. and K. Ohnishi (2015a). On the explicit robust force control via disturbance observer. *IEEE Transactions on Industrial Electronics 62*(3), 1581–1589.

Sariyildiz, E. and K. Ohnishi (2015b). Stability and robustness of disturbance-observer-based motion control systems. *IEEE Transactions on Industrial Electronics 62*(1), 414–422.

Schrijver, E. and J. v. Dijk (2002). Disturbance observers for rigid mechanical systems: Equivalence, stability, and design. *Journal of Dynamic Systems, Measurement, and Control 124*(4), 539–548.

Sharifzadeh, M., A. Arian, A. Salimi, M. T. Masouleh, and A. Kalhor (2017). An experimental study on the direct & indirect dynamic identification of an over-constrained 3-DOF decoupled parallel mechanism. *Mechanism and Machine Theory 116*, 178–202.

Sheng, P., M. Srinivasan, and S. Kobayashi (1995). Multi-objective process planning in environmentally conscious manufacturing: a feature-based approach. *CIRP Annals-Manufacturing Technology 44*(1), 433–437.

Shi, P., E.-K. Boukas, and R. K. Agarwal (1999). Control of Markovian jump discrete-time systems with norm bounded uncertainty and unknown delay. *IEEE Transactions on Automatic Control 44*(11), 2139–2144.

Silva, E. I. and D. A. Erraz (2006). An LQR based MIMO PID controller synthesis method for unconstrained Lagrangian mechanical systems. In *Proceedings of the 45th IEEE Conference on Decision and Control*, pp. 6593–6598.

Stearns, H., S. Mishra, and M. Tomizuka (2008). Iterative tuning of feedforward controller with force ripple compensation for wafer stage.

In *Proceedings of the 10th IEEE International Workshop on Advanced Motion Control*, pp. 234–239.

Sullivan, G. A., K. Suiter, and D. R. Huston (2000). Evaluation of gantry control strategies for X-ray steppers. In *Proceedings of the SPIE's 7th Annual International Symposium on Smart Structures and Materials*, pp. 597–606.

Sussmann, H. J. and J. C. Willems (1997). 300 years of optimal control: from the brachystochrone to the maximum principle. *IEEE Control Systems 17*(3), 32–44.

Tan, K. K., T. H. Lee, and S. Huang (2007). *Precision Motion Control: Design and Implementation.* London: Springer-Verlag.

Tan, K. K., X. Li, S.-L. Chen, C. S. Teo, and T. H. Lee (2019). Disturbance compensation by reference profile alteration with application to tray indexing. *IEEE Transactions on Industrial Electronics 66*(12), 9406–9416.

Tan, K. K., W. Liang, S. Huang, L. P. Pham, S.-L. Chen, C. W. Gan, and H. Y. Lim (2015). Precision control of piezoelectric ultrasonic motor for myringotomy with tube insertion. *Journal of Dynamic Systems, Measurement, and Control 137*(6), 064504.

Tan, K. K., S. Y. Lim, S. Huang, H. F. Dou, and T. S. Giam (2004). Coordinated motion control of moving gantry stages for precision applications based on an observer-augmented composite controller. *IEEE Transactions on Control Systems Technology 12*(6), 984–991.

Tan, K. K., Q.-G. Wang, and C. C. Hang (2012). *Advances in PID Control.* London: Springer-Verlag.

Tehrani, E. S., K. Jalaleddini, and R. E. Kearney (2017). Ankle joint intrinsic dynamics is more complex than a mass-spring-damper model. *IEEE Transactions on Neural Systems and Rehabilitation Engineering 25*(9), 1568–1580.

Teo, C. S., K. K. Tan, S. Y. Lim, S. Huang, and A. Tay (2007). Dynamic modeling and adaptive control of a H-type gantry stage. *Mechatronics 17*(7), 361–367.

Teo, T. J., G. Yang, and I.-M. Chen (2014). A large deflection and high payload flexure-based parallel manipulator for UV nanoimprint lithography: Part I. Modeling and analyses. *Precision Engineering 38*(4), 861–871.

Teo, T. J., H. Zhu, S.-L. Chen, G. Yang, and C. K. Pang (2016). Principle and modeling of a novel moving coil linear-rotary electromagnetic actuator. *IEEE Transactions on Industrial Electronics 63*(11), 6930–6940.

Tsitsiklis, J. and M. Athans (1985). On the complexity of decentralized decision making and detection problems. *IEEE Transactions on Automatic Control 30*(5), 440–446.

Utkin, V., J. Guldner, and J. Shi (2009). *Sliding Mode Control in Electro-Mechanical Systems.* Boca Raton: CRC Press.

van der Meulen, S. H., R. L. Tousain, and O. H. Bosgra (2008). Fixed structure feedforward controller design exploiting iterative trials: Application to a wafer stage and a desktop printer. *Journal of Dynamic Systems, Measurement, and Control 130*(5), 051006.

Wan, F. (1995). *Introduction to the Calculus of Variations and Its Applications.* London: Chapman & Hall Mathematics.

Wang, C.-C. and M. Tomizuka (2004). Design of robustly stable disturbance observers based on closed loop consideration using H_∞ optimization and its applications to motion control systems. In *Proceedings of 2004 American Control Conference*, Volume 4, pp. 3764–3769.

Wang, Q.-G., T. H. Lee, H. W. Fung, Q. Bi, and Y. Zhang (1999). PID tuning for improved performance. *IEEE Transactions on Control Systems Technology 7*(4), 457–465.

Wang, Y., L. Xie, and C. E. de Souza (1992). Robust control of a class of uncertain nonlinear systems. *Systems & Control Letters 19*(2), 139–149.

Wang, Y.-S., N. Matni, and J. C. Doyle (2018). Separable and localized system-level synthesis for large-scale systems. *IEEE Transactions on Automatic Control 63*(12), 4234–4249.

White, M. T., M. Tomizuka, and C. Smith (2000). Improved track following in magnetic disk drives using a disturbance observer. *IEEE/ASME Transactions on Mechatronics 5*(1), 3–11.

Wilamowski, B. and J. D. Irwin (2011). *Control and Mechatronics.* Boca Raton: CRC Press.

Xiao, S. and Y. Li (2013). Optimal design, fabrication, and control of an XY micropositioning stage driven by electromagnetic actuators. *IEEE Transactions on Industrial Electronics 60*(10), 4613–4626.

Xie, L. (1996). Output feedback H_∞ control of systems with parameter uncertainty. *International Journal of Control 63*(4), 741–750.

Xie, L. and E. de Souza Carlos (1992). Robust H_∞ control for linear systems with norm-bounded time-varying uncertainty. *IEEE Transactions on Automatic Control 37*(8), 1188–1191.

Xu, J.-X. and Y. Tan (2003). *Linear and Nonlinear Iterative Learning Control.* Berlin: Springer-Verlag.

Yan, M.-T. and Y.-J. Shiu (2008). Theory and application of a combined feedback–feedforward control and disturbance observer in linear motor drive wire-EDM machines. *International Journal of Machine Tools and Manufacture 48*(3), 388–401.

Yang, J., W. X. Zheng, S. Li, B. Wu, and M. Cheng (2015). Design of a prediction-accuracy-enhanced continuous-time MPC for disturbed systems via a disturbance observer. *IEEE Transactions on Industrial Electronics 62*(9), 5807–5816.

Yao, B., C. Hu, and Q. Wang (2012). An orthogonal global task coordinate frame for contouring control of biaxial systems. *IEEE/ASME Transactions on Mechatronics 17*(4), 622–634.

Yao, W., L. Jiang, J. Fang, J. Wen, and S. Cheng (2014). Decentralized nonlinear optimal predictive excitation control for multi-machine power systems. *International Journal of Electrical Power & Energy Systems 55*, 620–627.

Yao, W.-S. (2015). Modeling and synchronous control of dual mechanically coupled linear servo system. *Journal of Dynamic Systems, Measurement, and Control 137*(4), 041009.

Yin, S., X. Li, H. Gao, and O. Kaynak (2015). Data-based techniques focused on modern industry: an overview. *IEEE Transactions on Industrial Electronics 62*(1), 657–667.

Yuan, M., Z. Chen, B. Yao, and X. Zhu (2017). Time optimal contouring control of industrial biaxial gantry: A highly efficient analytical solution of trajectory planning. *IEEE/ASME Transactions on Mechatronics 22*(1), 247–257.

Yun, J. N. and J. Su (2013). Design of a disturbance observer for a two-link manipulator with flexible joints. *IEEE Transactions on Control Systems Technology 22*(2), 809–815.

Yun, J. N., J. Su, Y. I. Kim, and Y. C. Kim (2013). Robust disturbance observer for two-inertia system. *IEEE Transactions on Industrial Electronics 60*(7), 2700–2710.

Zhang, G., J. Chen, and Z. Li (2008). Analysis and design of H_∞ robust disturbance observer based on LMI. In *Proceedings of the 7th World Congress on Intelligent Control and Automation*, pp. 4697–4701.

Zhao, Z., Y. Liu, W. He, and F. Luo (2016). Adaptive boundary control of an axially moving belt system with high acceleration/deceleration. *IET Control Theory Applications 10*(11), 1299–1306.

Zhao, Z.-Y., M. Tomizuka, and S. Isaka (1993). Fuzzy gain scheduling of PID controllers. *IEEE Transactions on Systems, Man, and Cybernetics 23*(5), 1392–1398.

Zhou, K. and P. P. Khargonekar (1988). Robust stabilization of linear systems with norm-bounded time-varying uncertainty. *Systems & Control Letters 10*(1), 17–20.

Zhu, H., C. K. Pang, and T. J. Teo (2016). Integrated servo-mechanical design of a fine stage for a coarse/fine dual-stage positioning system. *IEEE/ASME Transactions on Mechatronics 21*(1), 329–338.

Zhu, H., C. K. Pang, and T. J. Teo (2017). A flexure-based parallel actuation dual-stage system for large-stroke nanopositioning. *IEEE Transactions on Industrial Electronics 64*(7), 5553–5563.

Zhu, H., T. J. Teo, and C. K. Pang (2017). Design and modeling of a six-degree-of-freedom magnetically levitated positioner using square coils and 1-D Halbach arrays. *IEEE Transactions on Industrial Electronics 64*(1), 440–450.

Index